LASER MATERIAL PROCESSING

William M. Steen

Laser Material Processing

With 206 Figures

Springer-Verlag
London   Berlin   Heidelberg   New York
Paris   Tokyo   Hong Kong
Barcelona   Budapest

William M. Steen
James Bibby Professor of Engineering Manufacture, Mechanical
Engineering Department, University of Liverpool,
Liverpool L69 3BX, UK

ISBN 3-540-19670-6 Springer-Verlag Berlin Heidelberg New York
ISBN 0-387-19670-6 Springer-Verlag New York Berlin Heidelberg

British Library Cataloguing in Publication Data
Steen, W. M.
Laser material processing.
I. Title
620.112
ISBN 3540196706

Library of Congress Cataloging-in-Publication Data
Steen, William M., 1933–
Laser material processing / by William M. Steen.
    p.   cm.
ISBN 0-387-19670-6
1. Lasers—Industrial applications.   I. Title.
TA 1677.S73   1991
621.36'6—dc20   91-31113
                                                         CIP

Printed and bound by Athenæum Press Ltd., Gateshead, Tyne & Wear.
69/3830 54321   Printed on acid-free paper

To Margaret, Pip and Mim

The stimuli for this emission

Acknowledgements

The author would like to acknowledge the support he has gained from the enthusiasm of the many students who have passed through the research schools he has led at Imperial College and now Liverpool. Many of these now run their own laser businesses, teach the subject or have found other ways of making money from laser material processing. This is one of the greatest pleasures an academic can have.

In help with this book he is particularly grateful to Dr. W. O'Neill for his help in editing and correcting chapters 1 & 2; to Prof D.Hall for guidance on brightness; to Dr. R. Akhter for his help in editing and assembling the illustrations for chapter 4; to Miss F. Fellowes for her help in assembling and editing chapter 6 and the cover picture, which is of silica powder blowing into a laser beam during the FOCAL process which she helped to invent; to Dr. L. Li for his help in editing and illustrating chapter 7; to Mrs. H. Garnet for helping with the typing of this Camera Ready Copy (CRC), and last but not least to his wife, Margaret, who created an ambience in which to live and complete this work.

Contents

Contents

To Margaret, Pip and Mim

The stimuli for this emission

Acknowledgements

The author would like to acknowledge the support he has gained from the enthusiasm of the many students who have passed through the research schools he has led at Imperial College and now Liverpool. Many of these now run their own laser businesses, teach the subject or have found other ways of making money from laser material processing. This is one of the greatest pleasures an academic can have.

In help with this book he is particularly grateful to Dr. W. O'Neill for his help in editing and correcting chapters 1 & 2; to Prof D.Hall for guidance on brightness; to Dr. R. Akhter for his help in editing and assembling the illustrations for chapter 4; to Miss F. Fellowes for her help in assembling and editing chapter 6 and the cover picture, which is of silica powder blowing into a laser beam during the FOCAL process which she helped to invent; to Dr. L. Li for his help in editing and illustrating chapter 7; to Mrs. H. Garnet for helping with the typing of this Camera Ready Copy (CRC), and last but not least to his wife, Margaret, who created an ambience in which to live and complete this work.

Contents

Prologue

Many hands make light work"
Erasmus Adages II iii 95 c1330

It has been true throughout history that every time mankind has mastered a new form of energy there has been a significant, if not massive, step forward in our quality of life. Due to the discovery of the laser in 1960 optical energy in large quantities and in a controlled form is now available as a new form of energy for the civilised world. So it is reasonable to have great expectations.

Consider the analogy with other forms of energy. The start of civilisation is identified with the ability to make tools - by the application of mechanical energy. The lower paleolithic of some 1.75M yrs ago produced ancient crude stone or bone tools. Finer stone tools were produced through the middle and upper paleolithic period to reach pinnacles of excellence over a development period of around 1.5M yrs. We might grumble about technology transfer being slow today but they really had an argument then! This simple technology based on the application of mechanical energy caused a major change in our quality of life - it took us out of the trees and converted us from animals to human beings.

Centuries later the control of chemical energy in the form of organised fires and convectively blown furnaces was achieved and the bronze (around 6500 B.C.) and iron ages (around 1500 B.C.) resulted in superior tools. Due to the increased productivity of agriculture using these superior tools and the improved security afforded from swords and chariots, stable political groupings formed, the Greek and Roman empires were born, the arts flourished and again a major step forward in our quality of life resulted.

The ability to harness wind energy in the form of windmills started industrialisation, while, in the form of sailing ships, it opened up the world to international trade. The great navigators discovered new worlds by applying wind energy with the help of the now sophisticated product of mechanical energy, the chronometer.

In 1701 Newcomen built the first working steam engine for pumping water at Dudley Castle in Chester, UK. By 1790 the Industrial Revolution was in full swing with steam engines doing the back breaking work of previous ages. The quality and the speed of life both increased.

In 1835 Faraday invented the dynamo and electricity became available in controllable form and in large quantities. The electric motor is the heart of many domestic machines and industrial plant. Arc welding, electric heating, radio, TV (there is a slight overlap with electromagnetic or optical energy here), telephone, lighting, computers and refrigerators are more examples of the dramatic effect the mastery of electrical energy has had on our quality of life.

Nikolaus Otto and Eugen Langen, working with the designs of Alphonse Beau de Rochas, started production of the first four stroke internal combustion engine in Deutz, Germany, in 1867 and so found a new way of harnessing chemical energy from oil and petrol. Personalised transport and flight became a reality. Although some argue this has brought no advance in our quality of life there is hardly a soul who would do without them. Swift travel has begun to make a true world community. International trade allows fresh vegetables all the year around, and the benefits of many cultures can be shared.

Nuclear energy became available when Enrico Fermi built the first atomic pile which went critical on the 2nd Dec 1942 in the squash court at the University of Chicago, USA. Atomic energy has been used directly only as a bomb or for medical radiation treatment. As such it has altered world politics. It has questioned the wisdom of settling arguments by fighting and so far has thus resulted in peace between the superpowers - though there may be a problem with the others. The current direct application of atomic power is only as a heating system in power stations, a form of coal substitute. Thus, I feel, that a further invention is needed before we have truly mastered this form of energy.

In 1960 Maiman [1] invented the first working ruby laser shown in Fig. 1. It was not a surprise but the result of considerable investment following Einstein's paper in 1917 [2] in which he showed that lasing action should be possible. With the great stories of H.G.Wells, "War of the Worlds" to guide them the military soon realised that a death ray would be handy on any battlefield. There resulted an avalanche of

Fig. 1. Theodore Maiman and his first ruby laser.

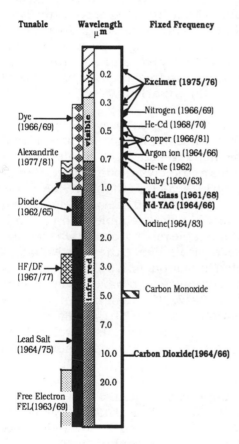

Fig. 2. Range of wavelengths for current commercial lasers. First date is date of discovery, the second is of commercialisation (4).

research funding - the only time I have heard of one laboratory requesting a grant for laser development and being awarded twice as much! However Maiman won this race by a few months with his solid state ruby laser. In the months and years following it seemed that almost anything could be made to lase. Fig. 2. shows the wavelength bands covered by commercial lasers of today.

In material processing the laser must be reasonably powerful and this reduces the number to only a few - essentially the CO_2, the Nd-YAG or Nd-glass and the Excimer lasers.

The CO_2 laser was invented by C.K.N. Patel in 1964 working at the Bell Laboratories (4). His first laser used pure CO_2 and produced 1mW of power with an efficiency of 0.0001%. By adding nitrogen the power improved to 200mW and when helium was added the power jumped to 100W with an efficiency of 6% - all this within a year! Today all CO_2 lasers have a gas mixture of approximately $CO_2:N_2:He::0.8:1:7$. The commercial potential for this sort of laser was immediately perceived and Spectra-Physics worked on developing the technology from 1965. They stopped a year later but the team working on the laser went on to found their own company "Coherent". They marketed the first CO_2 laser at 100W in 1966 and a 250W version in 1968. The need to cool the gas was soon understood and methods of convectively cooling lasers were designed. In 1969 GM Delco installed 3 lasers for industrial work. AVCO came out in the early 1970s with a 15kW CO_2 transverse flow laser. It was an anachronism and the market was not ready for such a powerful laser; but this laser gave an insight into the potential for material processing even though the mode was poor and so the focus was never very fine. Today the CO_2 laser has developed into the work horse for material processing with slow flow, fast axial flow and transverse flow lasers operating at powers up to 25kW or even 100kW for military funded laboratories. High powered pulsed CO_2 lasers have also been developed. The TEA lasers (Transverse Excitation Atmospheric pressure) have megawatts of power operating in a pulsed mode with up to 10J/pulse and 1ms pulse lengths. Another growth direction has been into sealed low power units, usually waveguide lasers, for medical and guidance uses which are now available with up to 100W or so output with a lifetime of around 4000 hours. CO_2 lasers fitted with a suitable grating as a rear mirror have an output which can be a single spectral line. This spectral line can be tuned between 8 and 11μm wavebands. The tuning can be continuous if the pressure of the tube is increased. This has allowed growth into the optical and communications markets.

The Nd-YAG laser was invented in the Bell Laboratories in 1964. Quantronix, Holobeam, Control Laser (who later bought Holobeam) and Coherent were quick to enter the market since the application of resistance trimming for the electronics market was soon appreciated as a large potential market. But the market was less than 1M$ for many years since the lasers were of poor quality. In 1976 the market changed abruptly when Quanta-Ray introduced the first reliable high performance YAG laser. The design used an unstable resonator and was robust. It gave 1J pulses at half the price of its rivals. Since then the Nd lasers have progressed to have sophisticated individual pulse shaping such as those developed by JK Laser (now Lumonics who have since become Sumitomo) with powers ranging up to 3kW. The host material for the Nd has also developed into several varieties: still the Yttrium Aluminium Garnet (YAG) crystal, but now a variety of glass materials as well as YLF, Yttrium Lanthanum Fluoride, YAP, Yttrium Aluminium Phosphide and others (Fig. 3).

The excimer laser, working on the improbable chemistry of noble gas halides, e.g. KrF, was invented in 1975/76. These ultraviolet lasers seemed to have potential for military purposes due to the low reflectivity of short wavelength radiation. Considerable money has been put into developing them and as a result the material processing industry now has some robust ultra violet lasers capable of "cold cutting" with immense prospects for the electronic industry, microlithography, weird chemistries and 9ns suntans!

As a result of this activity we now have optical energy in a controllable form. The question is: Will the mastery of this form of energy also give a massive boost to our standard of living?

In the previous examples there were people who got hurt and those who did well from these changes. That is the law of natural selection which has been the underlying theme to all changes in this world. It is not something to control so much as something to note as a lesson from life. It is also the hard reality associated with progress. Our politicians and ourselves might try to mitigate the suffering, yes; but fight the changes, no! The application of optical energy will be no exception - why should it be? Is it possible to stop the clock and pretend something has not been invented? The winners in the past have always been those who see change as an opportunity, not a threat. I hope the reader is of that mind and that this book may help open new opportunities for him or her.

References

1. Maiman.T.H. "Stimulated Optical Radiation in Ruby" Nature Aug 6th 1960.
2. Einstein.A. Phys Z, 18 121 1917.
3. Klauminzer.G.K. "Twenty years of Commercial Lasers-A capsule History" Laser Focus/Electrooptics Dec 1984 pp54-60.
4. Patel.C.K.N. 1964 Phys Rev 136 A1187.

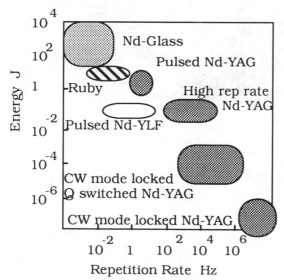

Fig. 3 Operating ranges of commercial solid state lasers.

"Perhaps when you have finished messing around, you could give me a hand with these wheels."

Chapter 1

Background and General Applications

"To have begun is half the job: be bold and be sensible"
Horace 65-8B.C. Epistles I ii 40.

1.1. How the Laser Works

1.1.1. Construction

1.1.1.1. Overall Design: The basic laser consists of two mirrors which are placed parallel to each other to form an optical oscillator, that is a chamber in which light would oscillate back and forth between the mirrors for ever, if not prevented by some mechanism such as absorption. Between the mirrors is an active medium which is capable of amplifying the light oscillations by the mechanism of Stimulated Emission (the process after which the laser is named - **L**ight **A**mplification by the **S**timulated **E**mission of **R**adiation). We will return to this process in a moment. There is also some system for pumping the active medium so that it has the energy to become active. This is usually a DC or RF power supply, for CO_2 and He/Ne lasers, or a focussed pulse of light, for the Nd-YAG laser. It may, however, be a chemical reaction, as with the iodine laser. The optical arrangement is shown in Fig 1.1a,b,c. One of the two mirrors is partially transparent to allow some of the oscillating power to emerge as the operating beam. The other mirror is totally reflecting, to the best that can be achieved (99.999% or some such figure). This mirror is also usually curved to reduce the diffraction losses of the oscillating power and also to make it possible to align the mirrors without undue difficulty - this would be the case if both mirrors were flat. The design of the laser cavity hinges on the length of the cavity and the shape of these mirrors, including any others in a folded system.

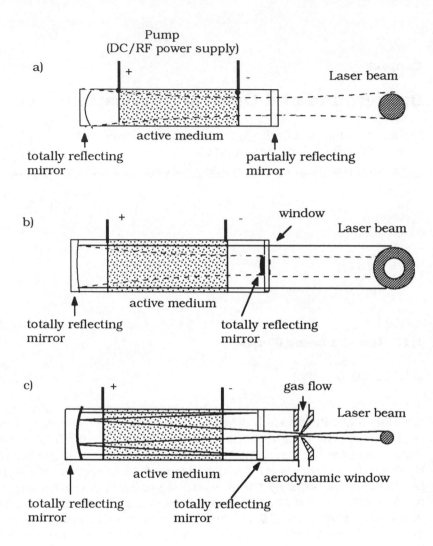

Fig. 1.1. Basic construction of a laser cavity.
a) Stable; b) Unstable; c) Stable cavity with aerodynamic window.

1.1.1.2. Cavity Mirror Design:

Kogelnik and Li (1) wrote one of the fundamental papers on cavity design. They showed by geometric arguments that the mirror curvatures at either end of the cavity could only fall within certain values or the cavity would become "unstable" by losing the power around the edge of the output mirror. Cavities can be identified as "stable" or "unstable" depending upon whether they make the oscillating beam converge into the cavity or spread out from the cavity. Most lasers, up to 2kW, use stable cavity designs (Fig. 1.1a) because it is safe to transmit that level of power through the output mirror without risk of breakage. The output mirror, being partially transparent to 10.6μm infrared radiation, is made of ZnSe (Zinc Selenide), GaAs (Gallium Arsenide), or CdTe (Cadmium Telluride) for CO_2 lasers and BK7 fused silica glass for a YAG laser. In all cases it is carefully coated to give the required level of reflection into the cavity (typically a CO_2 laser output window would have a 35% reflectivity for feedback into the cavity). Breakage of the output window is serious from the implosion aspect and the problems associated with shut down in production and cost. Thus for higher powered lasers it is not uncommon to find the cavity is designed as an unstable cavity (Fig. 1.1b) taking the power from around the edge of the output mirror, which is in this case a totally reflecting metal optic. The larger ring shaped beam thus passed means reduced power density on the window sealing the vacuum chamber in which the cavity sits. An alternative is to have an aerodynamic window, in which a venturi arrangement ensures that the vacuum is held while the beam passes through the high velocity low pressure gas to atmosphere (Fig. 1.1c). The shape of a stable beam is determined by the shape of the output aperture; while the shape of the unstable beam is the same as the edge of the output reflecting mirror. It is usually round or ring shaped but may be square as with some excimer lasers and slab lasers.

1.1.1.3. Cavity length: The length of the cavity to the width of the output aperture determines the number of off axis modes which can oscillate within the cavity between the two mirrors. This number is described in the Fresnel number, $N = a^2/\lambda L$, a dimensionless group, where a = radius of the output aperture, L = length of the cavity, and λ = wavelength of laser radiation. The Fresnel number, N, equals the number of fringes which would be seen at the output aperture if the back end mirror was uniformly illuminated. Thus a low Fresnel number gives a low order mode. Off axis oscillations are lost by diffraction and hence will not occur in an amplifying cavity. Table 1.1 lists some of the Fresnel numbers associated with some current industrial lasers. A higher Fresnel number cavity may be controlled to give lower order beam modes by controlling the mirror design; an example is the flexible mirror design of the Laser Ecosse AF5 laser. This laser can be engineered to give any mode from TEM00 to a multi mode beam.

The off axis modes describe the Transverse Electromagnetic Mode (TEM) structure of the power distribution across the beam which is essentially a standing wave across the beam. This mode structure is discussed in the next chapter (Section 2.5.3). The mode coming from a given laser can be modified by inserting apertures into the cavity, to alter the cavity

Table 1.1	Dimensions of Typical Industrial Lasers			
Model	Cavity radius mm	Cavity length m	Fresnel number	Mode
Laser Ecosse MF400	3.5	14.4	0.8	TEM01*/TEM00
Electrox	9	3.4	2.2	TEM01*
PRC1500	9	3.4	2.2	TEM01*
Control 2kW	17.5	6	4.8	Low
PRC3000	11	5.17	2.2	TEM01*/Low
Laser Ecosse AF5	10	15	0.6	TEM00 controllable

Fresnel number. This will, of course, reduce the power output but may give a more easily focussed beam.

1.1.2. Stimulated Emission Phenomenon

The mystical part of a laser is that it works at all. This is entirely due to the stimulated emission phenomenon. It was predicted by Einstein in 1917 (2) using a mathematical argument. It was thus not a phenomenon of everyday experience.

Consider the carbon dioxide molecule. It can take on various energy states depending upon some form of vibration or rotation. These states are quantised, that is they can only exist at particular energy levels or not at all. The basic energy network possible with carbon dioxide is shown in Fig. 1.2. The gas mixture in a carbon dioxide laser is subject to an electric discharge causing the low pressure gas (usually around 35-50 torr) to form a plasma. In the plasma the molecules take up various excited states as expected from the Boltzmann distribution ($n_i = Ce^{-E/kT}$; where n_i = number of molecules in energy state i; E = energy of state i; k = Boltzmann constant; T = Absolute temperature; C = constant). Some will be in the upper state (00^01) which represents an asymmetric oscillation mode. By chance this molecule may lose its energy by collision with the walls of the cavity or by spontaneous emission. Through spontaneous emission the state falls to the symmetric oscillation mode (10^00) and a photon of light of wavelength $10.6\mu m$ is emitted travelling in any direction dictated by chance. One of these photons, again by chance, will be travelling down the optic axis of the cavity and will start oscillating between the mirrors. During this time it can be absorbed by a molecule in the (10^00) state, or it can be diffracted

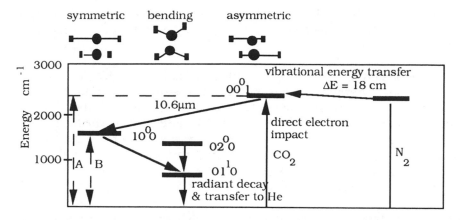

Fig. 1.2. Energy levels of the carbon dioxide molecule.

out of the system or it will strike a molecule which is already excited - in the (00^01) state. At this moment it will stimulate that excited molecule to release its energy and fall to a lower energy state and thus emit another photon of identical wavelength, travelling in exactly the same direction and with the same phase. One can imagine the incident photon shaking the energy free by some form of resonance. The two photons travelling in the same direction with the same phase now sweep back and forth within the cavity generating more photons from other excited molecules. The excited state becomes depleted and so by the Boltzmann distribution more and more of the energy is passed into that state giving a satisfactory conversion of electrical energy into the upper state. It is necessary for another condition to hold, and that is that the excited state which lases should be slow to undergo spontaneous emission and the lower state should be faster. This allows an inversion of the population of excited species to exist and thus makes a medium which is more available for the stimulated emission process (amplification) than for absorption.

In actual fact if that were the whole story then the CO_2 laser would not be special nor the most powerful laser available today. The CO_2 laser is helped by a quirk of nature whereby nitrogen, which can only oscillate in one way, being made of two atoms, has an energy gap between the different quanta of oscillation which is within a few Hertz of that required to take cold CO_2 and put it into the asymmetric oscillation mode - the upper laser level (00^01). Thus by collision with excited nitrogen, cold CO_2 can be made excited. One of Patel's first experiments in 1964 used this phenomenon. He excited the N_2 and passed it into the CO_2 gas and made it lase even though the CO_2 was not in the plasma area. The only way the nitrogen can lose its energy is by collision with the tube walls or a molecule which will absorb that energy. Thus its lifetime is long and the efficiency of the CO_2 laser is high (15-20%); but it is a function of the gas temperature, since only cold CO_2 will undergo this reaction with nitrogen.

Thus the design of a CO_2 laser is built around the requirement to have cool CO_2 gas. Firstly, the gas mixture in the laser is around 78% He for good conduction and stabilisation of the plasma, 13% N_2 for this coupling effect and 10% CO_2 to do the work. The efficiency of the laser is not a strong function of the gas composition except for certain impurities. Secondly the gas is cooled by conduction through the walls for slow flow lasers (SF lasers) or by convection in the fast axial flow (FAF and AF lasers) and transverse flow lasers (TF lasers).

Many materials can be made to show this stimulated emission phenomenon, only a few with significant power capability. The main lasers used for material processing are: CO_2, CO, Nd-YAG, Nd-Glass and Excimer (e.g. KrF, ArF, XeCl).

Carbon Dioxide: The energy diagram for a CO_2 laser has just been discussed and is shown in Fig. 1.2. The quantum efficiency of such lasers is 45% (that is the ratio A/B, Fig. 1.2). So far the operating efficiency achieved is around 15-20% for electric discharge to optical power but only around 12% for wall plug efficiency (Optical energy out/ Total electrical energy into the system). Typical values for the wall plug efficiencies for these main lasers are given in Table 1.2.

Table 1.2	Efficiency of main types of industrial lasers	
Type	Wavelength μm	Wall Plug Efficiency %
Carbon Dioxide	10.6	12
Carbon Monoxide	5.4	8
Nd-YAG	1.06	0.4
Nd-Glass	1.06	0.1
Excimer (KrF)	0.249	2

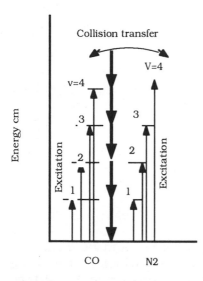

Fig. 1.3. Vibrational energy levels for the CO molecule.

Carbon Monoxide: The CO laser, whose energy diagram is shown in Fig. 1.3, has the advantage of a quantum efficiency of near 100% and thus promises to have a wall plug efficiency of twice that of the CO_2 laser, though this is seriously reduced by the cooling requirements. For high powered lasers this could be significant since a 100kW CO_2 laser would require a power supply of 0.83MW at least. This is getting near to being a small power station. Half that value is more practical. However the CO laser currently operates best at very low temperatures around 150K (liquid nitrogen temperatures) and requires extensive power for refrigeration, which may affect this potential efficiency advantage. Currently operating efficiencies of around 19% have been reported but these are reduced to nearer 8% when the power for cooling is considered.

Neodymium-Yttrium Aluminium Garnet: The Nd-YAG laser has an energy diagram shown in Fig. 1.4. The quantum efficiency is 40%. The operating efficiency is low since the pumping is done with a broad band illumination of which only a proportion of the radiation is able to excite the neodymium atoms in the crystal. It thus lacks the natural coupling to N_2 of the CO_2 laser. This means that considerable energy has to be pumped into the crystal rod giving a serious cooling problem. For this reason the YAG laser is currently limited to around 400W/rod before serious beam distortion due to thermal effects occurs. The total power of a system may be increased by the use of oscillator-amplifier arrangements or the focussing of bundles of beams as with the Lumonics device shown in Fig. 1.5. Possibly the main advances will be made by devising a more effective pumping system using diode lasers of the appropriate frequency, to pump the neodymium with greater precision.

Neodymium-Glass: The Nd-Glass lasers have the same energy diagram for Nd as the YAG laser but the energy conversion is better in glass. However the cooling problems are more severe and so the Nd-glass lasers are confined to slow repetition rates, ~1Hz (see Fig. 3, Prologue). At higher repetition rates the beam divergence (or ease of focussing) becomes unacceptable for material processing. The beam from a glass laser is more spiked than that from a YAG as seen in Fig. 1.6. It is more prone to burst mode operation.

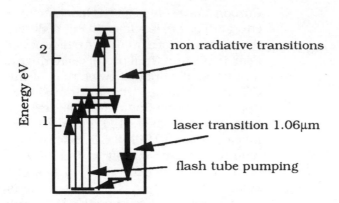

Fig. 1.4. Energy levels for Neodymium.

Excimer: The Excimer laser has an energy diagram shown in Fig. 1.7.
The name derives from the excited dimer molecules which are the lasing
species.   There are several gas mixtures used in an excimer laser; they
are shown in Table 1.3.

These lasers are slightly different in that the gain is so strong that they

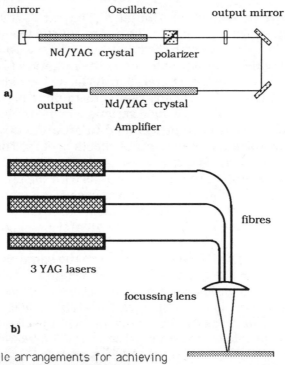

Fig. 1.5. Possible arrangements for achieving
powers of over 1kW with a Nd-YAG laser.
a) Oscillator/amplifier. b) Coupled fibres.

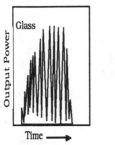
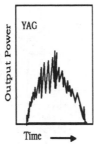
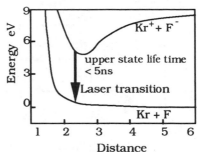

Fig. 1.6. The spiky nature of Nd- glass
pulses compared to Nd-YAG.

Fig 1.7. Energy levels for an
excimer laser.

do not need an oscillator. An electric discharge is generated, for example, in a gas mixture of Kr, F_2 and Ne at around four atmospheres. An excited dimer KrF^* is formed which undergoes the stimulated emission process. It generates ultra violet photons in a brief pulse for each discharge of the condenser bank into the gas mixture. The pulses are usually very short, around 20ns (a piece of light around 6m long!) but very powerful, typically around 35MW (the energy/pulse is thus 0.2J/pulse). Due to the lack of resonant oscillation the mode from these lasers is very poor. The process has more to do with amplified spontaneous emission than with laser oscillation.

Table 1.3

Range of wavelengths for different gas mixtures in an excimer laser.

Gas Mixture	Wavelength nm
KrF	248
ArF	193
XeF	354
KrCl	222
XeCl	308

1.2. Types of Industrial Lasers

1.2.1. Carbon Dioxide Lasers

The essence of the design of CO_2 lasers is the cooling of the carbon dioxide gas mixture.

1.2.1.1. Slow Flow Lasers (SF): With these lasers the cooling is through the walls of the cavity. The general arrangement is shown in Fig. 1.8. Typical operating figures are 20 1/min gas flow, 7 1/min coolant flow, 20°C temperature and gain of around 30-50W/m. The gain/m is relatively small and so these lasers are either very long as with the first powerful lasers at Essex University in the 1960s, which ran to some 70m in a straight line, or not very powerful - up to 2kW. In any case the cavity length will be long but may be subtly folded. These long cavities mean a low Fresnel number and hence a low order mode which is the most suitable form for focussing - see Chapter 2. The relatively smooth plasma formed in the slow flowing gases also ensures a good mode. These lasers are amongst the best for cutting due to this superior mode.

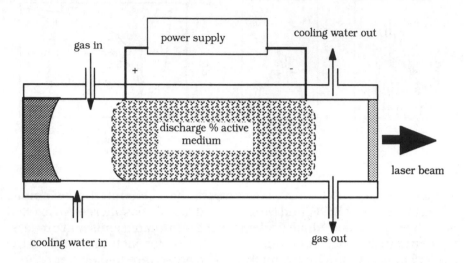

Fig. I.8. Basic construction of a slow flow(SF) laser.

1.2.1.2. Fast Axial Flow Lasers (FAF): FAF lasers achieve their cooling by convection of the gas through the discharge zone. The general arrangement is shown in Fig. 1.9. Typically the gases flow at 300-500 m/s through the discharge zone. Control of the gas mixture and avoidance of any leakage allows smooth plasmas to be produced. The axial nature of the flow, discharge and optical oscillation favours an axially symmetric power distribution in the beam. The cavity length is usually of a fairly low Fresnel number and so the beam mode is of a low order and thus more easily focussed to a small point. The gain is typically 500W/m and hence compact high power units have been made this way. Pumping can be by DC or RF discharge. 5kW lasers of Laser Ecosse and Trumpf are shown in Fig. 1.10.

1.2.1.3. Transverse Flow Lasers (TF): These lasers are once more con-vectively cooled but this time the flow is transverse to the discharge. Cooling is thus more effective and very compact high power lasers have been built this way. The general arrangement is shown in Fig. 1.11. The original AVCO 15kW laser was of this type in 1971. The current lasers include the UTRC 25kW Fig. 1.12, and the CL10, Fig. 1.13. The main disadvantage of these lasers lies in the lack of flow symmetry. For example, the gas enters the cavity cold and becomes heated as it traverses the lasing space. Thus the gain, which is a function of temperature, falls across the cavity and an asymmetric beam power results.The MLI and Trumpf TLF5000 lasers have a square cavities which attempt to smooth this effect. The design is shown in Fig. 1.14.

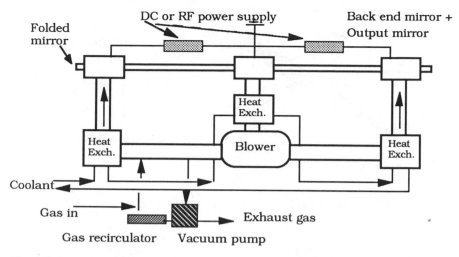

Fig. 1.9. General construction of a fast axial flow (FAF) laser.

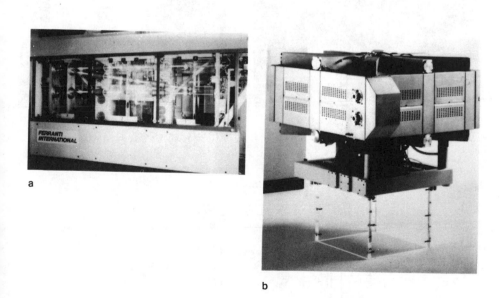

Fig. 1.10. Industrial FAF lasers: a) Laser Ecosse AF5, b) Trumpf TFL5000 Turbo.

1.2.1.4. Other Designs:

Various flow patterns on these general themes have been and are being explored. One was the Photon Sources Turbo Laser using a spiral flow which rotated in and out of the discharge zone.

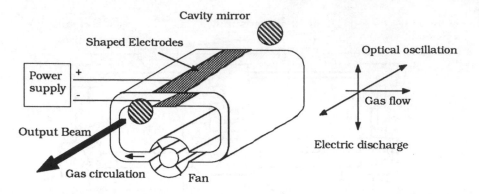

Fig. 1.11. General construction of a transverse flow (TF) laser.

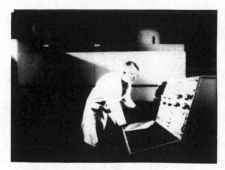

Fig. 1.12. United Technology
UTRC 25kW laser.

Fig. 1.13. Laser Ecosse 10kW
CL10 TF laser.

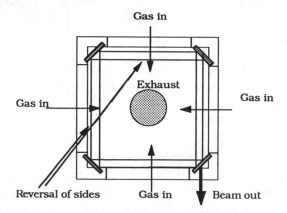

Fig. 1.14. The square cavity for more uniform amplification across the beam.

1.2.2. Carbon Monoxide Lasers

These lasers are not currently commercially available but the designs being considered are similar to those for a fast axial flow system with added cooling from liquid nitrogen or special refrigeration. Some designs include substantial cooling by pressure expansion (Joule Thomson cooling).

1.2.3. Solid State Lasers

The essence of the design of a solid state laser is how to get the pumping power into the laser block and cool the block, in such a way that it does not distort or break.

1.2.3.1. Nd-YAG lasers: The overall construction of a Nd-YAG laser is shown in Fig. 1.15. It consists of the standard cavity design with the active medium being neodymium in a Yttrium Aluminium Garnet (YAG) crystal rod mounted at one of the focii of an elliptical cavity. At the other focus is a krypton lamp. Also mounted in the optical cavity is an aperture for mode control, and possibly a Q switch for rapid shuttering of the cavity to generate fast pulses of power. Q switching is not possible with the powerful 400W lasers but is useful for resistance trimmer lasers rated at a few watts of power. In these cases the peak power per pulse will be around 10kW or so when Q switched (Q stands for "quality"). The rest of the laser seen in Fig. 1.15. consists of beam collimation and guidance systems which are mounted outside the cavity.

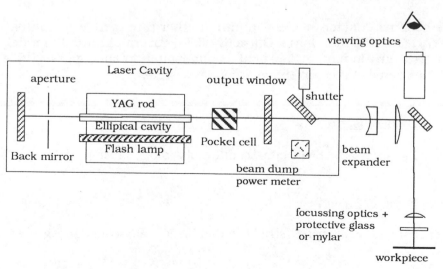

Fig. 1.15. General construction of a Nd-YAG laser.

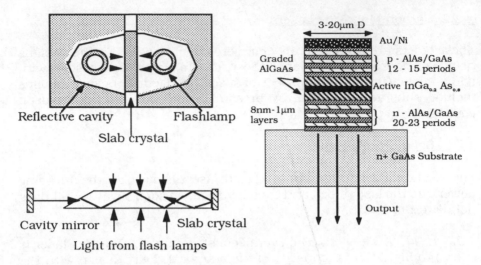

Fig. 1.16. The slab laser concept.   Fig. 1.17. Surface emitting diode
                                             laser (SEL).

A modification of this pattern is the introduction of slab or face pumped YAG crystals. These are costly but may have the ability to generate 1kW of power with sufficient cooling area to avoid distortion of the beam during operation. The first slab laser was invented in 1972 in Marshal Jones' group at GE, USA. This geometry reduces thermal lensing compared to the rod design allowing stronger pumping. A recent design is shown in Fig. 1.16 (3).

The host material for the neodymium or other rare earth element may be YAG (Yttrium Aluminium Garnet), YLF (Yttrium Lithium Fluoride), YAP (Yttrium Aluminium Phosphate) or phosphate or silica glass. Table 1.4 lists some of the current commercial systems.

1.2.4. Diode Lasers

Fig. 1.17 shows the basic construction of a diode laser.

1.2.5. Excimer Lasers

The construction of an industrial excimer laser is illustrated in Fig. 1.18.

Table 1.4	Wavelengths accessible with common solid state lasers							
Laser Type	Wavelength (μm)							
	0.1	0.2	0.3	0.4	0.6	0.8	1.0	2.0
Holmium-YAG							*	*
Erbium-Glass					*		*	
Nd-YAP		*	*	*			*	
Nd-YAG		***	*	* **	*		* *	
Nd-YLF		* *	*	*			*	
Nd-Silica Glass		*	*	*			*	
Nd-Phosphate Glass		*	*	*			*	
Ti-Sapphire			*	****	**	****	*	
Cr-Alexandrite			***	*	*	*		
Cr-Ruby			*		*			

* approximate region of principle wavelengths.

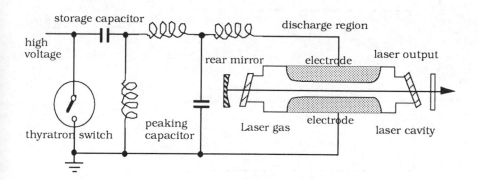

Fig. 1.18. Basic construction of an excimer laser.

1.3. Comparison Between Lasers

The comparison of these lasers can be made in several ways as illustrated in Fig. 1.19a,b,c.
a) Maximum power levels achieved for different wavelengths.
b) Capital cost shown for different types of laser.
c) Operating costs in £/W for different types of laser.

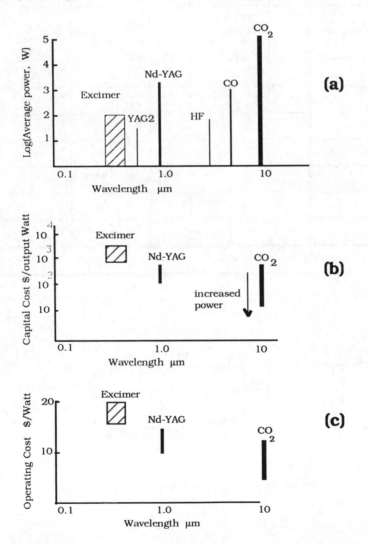

Fig. 1.19. Various comparisons between the principle material processing lasers;
a) Comparison of available power vs wavelength YAG2 = frequency doubled Nd:YAG.
b) Comparison of capital cost/output watt..
c) Comparison of operating cost/output watt.

1.4. Applications of Lasers

The laser was invented in 1960 and was soon dubbed as "a solution looking for a problem". So new was the tool that our thinking had not caught up with the possibilities. Today the story is distinctly different. Table 1.5 lists most of the areas of application. They fall into three groups: optical uses, power uses as in material processing and ultra power uses for atomic fusion. The range of applications is briefly discussed here only by way of background to material processing.

Table 1.5.	General applications of lasers.						
Application	Property of beam most used						Laser normally used
	Mono-Chromatic	Low Divergence	Coherent	High Power	Single Mode	Efficient	
Powerful Light		■		▓			He/Ne, Argon
Alignment	■	■					He/Ne
Measurement of Length	■	▓	■	▓	▓		He/Ne, Ruby, Nd-Glass
Pollution Detection	■						Dye, GaAs
Velocity Measurement	■						He/Ne, Nd-Glass
Holography	■	▓	■		■		All, Mainly Visible
Speckle Interferometry	■		■				He/Ne
Inspection	■		■	▓			He/Ne, Ruby
Analytical Technique		■					Nd-YAG
Recording	░	■					GaAs, GaAsP
Communications	■	■					He/Ne, GaAs, Iodine
Heat Source	▓		■	■	░	■	CO2, Nd-YAG or Glass Excimer
Medical	■		■	■			CO2, Ruby, Argon, Excimer
Printing	■			▓			He/Ne, Argon
Isotope Separation	■	░		■			Dye, Argon, Copper
Atomic Fusion	■	░		■			CO2, Nd-Glass
shade coding	■ Main property		▓ Second property		░ Third property		

24   Laser Material Processing

1.4.1. Powerful Light

The beam from a laser can have a low divergence and hence can be projected to make a bright spot, as with laser pointers for lecturing. By moving the spot, a pattern is retained in the eye generating a form of laser light show. Strange patterns can be made by rastering, the use of optical gratings, screwed up transparent paper and many other optical components, leading to the mind disorienting effects of a laser disco or laser light show.

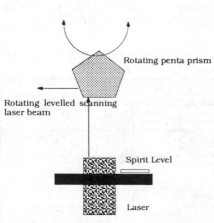

Rotating penta prism

Rotating levelled scanning laser beam

Spirit Level

Laser

Fig. 1.20. A laser level.

1.4.2. Alignment

Laser theodolites with automatic level read off are available. Tunnelling is now guided by lasers. The flexing of bridges, super tankers and the movement of glaciers is now recorded by laser alignment techniques. By sweeping the laser beam around, a line is marked out. If the sweeping mechanism is carefully levelled then a rapid technique for ground levelling, used by farmers and road builders, is instantly available. The beam is usually passed through a penta prism which reduces the need to carefully align the prism to the beam. This is then rotated about the beam axis to produce a level signal. The laser used is usually a visible, red, He/Ne laser (Fig. 1.20).

1.4.3. Measurement of Length

This can be done in several ways.

1.4.3.1. Interference: Using the coherent nature of a laser beam, whereby the beam is a continuous wave stream, it can be used as a form of ruler in an interferometer which can have very different path lengths for the two interfering beams. The basic design of the interferometer is shown in Fig. 1.21. 0.1μm accuracy for the positioning of machine tables has been achieved this way with high levels of repeatability (4). The arrangement is illustrated in Fig. 1.22. Distances up to the coherence length of the laser beam can be measured this way. This is often around 100m or so. *(The coherence length is the maximum path difference two rays can travel and yet be able to interfere constructively. This is similar to the length of a continuous wave stream from the laser. For a He/Ne laser it is around 70cm and for a CO_2 laser several 100m.)*

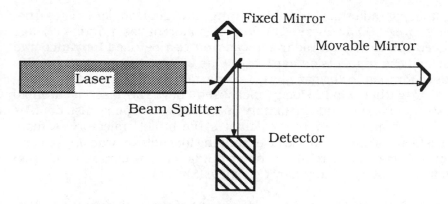

Fig. 1.21. A Michelson type interferometer for measuring length.

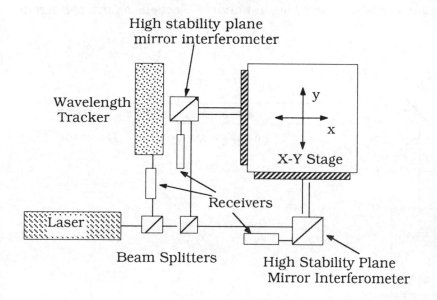

Fig. 1.22. A laser calibrated x/y table.

1.4.3.2. Time of Flight: Distance can be measured by the flight time of a short pulse. This is how it was found that the moon wobbled. It is how range finders work. Any distance of around a kilometre can only be measured this way if a laser is used. The accuracy is approx: ± 2cm, depending on the pulse length. Currently there is a handheld device operating at 72μW-and costing around $10,000 at 1990 prices - which is used by the customs to check the length of containers and lorries for finding false walls etc.(5). Laser radar(6)has been developed. TheFirefly

CO_2 Imaging radar has been able to range and get Doppler images from targets over 800km away. The use of coherent laser radar (7) has reduced the clutter on radar screens and can be used for range and velocity measurements on hard targets as well as direction. It can also be used for monitoring gas phase and particulate matter in the atmosphere to gather wind velocity measurements. The laser is used to substitute tracer rounds in military operations. The pulse also doubles as a ranger and a computer calculates the bullet trajectory. A more futuristic possibility is to use laser ranging for collision warning in cars, automatic speed control or even car guidance - in fact as a collision avoidance system for any moving article (8).

1.4.3.3. Occlusion Time: By measurement of the occlusion time of the scanning beam shown in Fig. 1.23, the width of a wire can be measured, while the wire is being made and travelling at speeds of a few km/s. The technique can also be used on stationary objects as a form of micrometer.

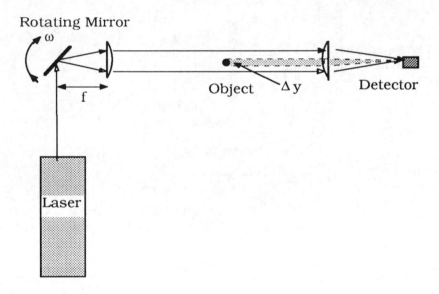

Fig. 1.23. A method of measuring thickness by the occlusion time.

The beam scans the object at a speed of $v = 2\omega f$; (Beam moves at twice mirror speed). Where ω = angular velocity in radians/s and f = focal length of lens in m (see Fig. 1.23). Thus the occlusion time, Δt, gives the dimension of the component as $\Delta y = v\Delta t = 2\omega f \Delta t$. The accuracy of this system depends on the beam size at the plane of the object.

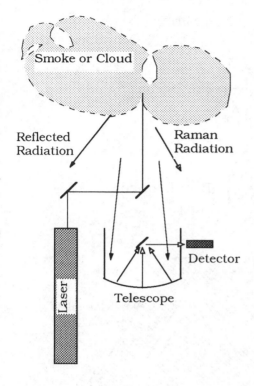

Fig. 1.24. The LIDAR system.

1.4.4. Pollution Detection

A burst of light of a particular frequency can not only measure the distance of an object but if the object is a cloud of steam or smoke it may excite the molecules of, say SO_2, in the cloud and as this decays a Raman spectrum is revealed which indicates the concentration of the pollutant in the cloud. This system, known as LIDAR (Light Detection and Ranging), is illustrated in Fig. 1.24. The laser can also be used to measure combustion products as in the UKAEA Harwell LISATEK 350 (9) which stands for Light Instrument for Sizing and Anemometry. Denver University have a device for measuring the CO emission from cars passing in single file at less than 45m.p.h. The device takes only 1.25s/reading (10).

1.4.5. Velocity Measurement

There are two types of laser velocity meters. The Laser Doppler Velocimeter measures the frequency shift in the radiation emitted and the radiation returning as illustrated in Fig. 1.25. It works best with a cooperative target but this is not strictly necessary. The pulse flight time also records the distance which is handy in the military context. These instruments can be used in measuring the high speed movement of drop hammers and other machine movements. A variation on this is to have a triangular laser as shown in Fig. 1.26. As the triangle is rotated so the optical path length differs for the light rotating in one direction compared to that in the other. The result is a beat frequency which can be detected on the detector mirror. This device is thus a gyroscope which cannot be toppled. It is currently fitted in many airliners and missiles.

The other meter is the Laser Doppler Anemometer (LDA) illustrated in Fig. 1.27. This instrument is now standard equipment for those studying fluid flow phenomena. The split beam is reunited at the point in a flowing stream to be analysed. At this point the coherent beam will form

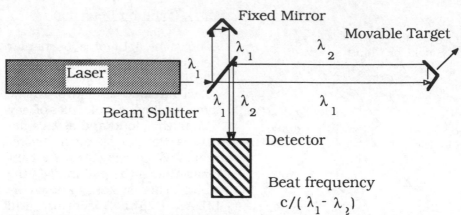

Fig. 1.25. The Laser Doppler Velocimeter.

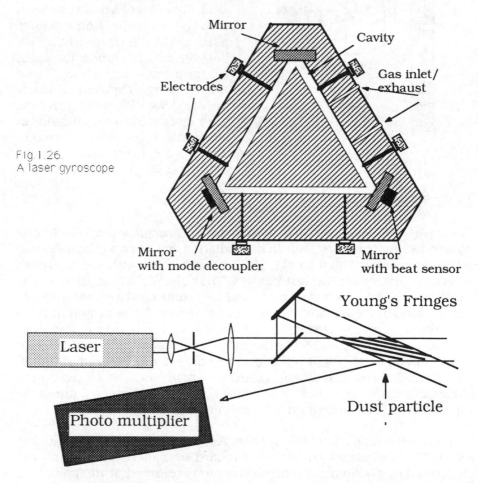

Fig. 1.26.
A laser gyroscope

Fig. 1.27. Principle of a Laser Doppler Anemometer (LDA).

Young's fringes, as illustrated in Fig. 1.27, where the phases of the two beams are in step or out of step. As particles flow through the fringes they will reflect the beam (the system usually needs some smoke or dust addition for a good signal). This reflected signal will be detected by a photomultiplier as a series of flashes. The frequencies in this signal can be analysed by fast Fourier transformation (FFT) and the velocity of the particles in the plane of the fringes analysed, including the velocity variation and thus turbulence. The whole analysis is done without interfering with the flow in any way and can be done remotely, for example to measure the flow within a diesel engine while it is operating. Since this method only measures one velocity vector, a two colour system may be used for measuring two dimensions simultaneously. By analysing the precise frequency of the return signal the Doppler shift would indicate the velocity in the direction along the beam axis. Thus this instrument is capable of measuring the velocity in all three dimensions simultaneously.

1.4.6. Holography

This is a true three dimensional form of photography which requires no lens in the camera! The arrangement for making a hologram ("*holo*" = whole image) is shown in Fig. 1.28. The film is thus exposed to the direct beam, which can be considered as a time marker and the reflected beam from the object, which gives data on the shape and illumination of the object. The interference pattern which results on the photographic plate thus scores the shape, illumination and time of arrival of the waves from the object all over the plate. By shining the light through the developed hologram, as in Fig. 1.29, the wave front is reconstructed as before. The definition is dependent on the grain size of the film. Special films have been developed for holography, for example the Denysik holograms with <5μm emulsion, and the subject has been advanced with the making of white light holograms. Holograms can be made using a light sensitive plastic which causes a change in the refractive index on exposure. These holograms do not require developing. Moving holograms are being developed, though currently the framing speed is a little slow at 2 frames/s (10).

1.4.6.1. Holographic Interferometry: This is achieved by developing a hologram of, for example a turbine blade, and then mounting the image from the hologram in exact alignment with the blade itself. The blade viewed through the hologram will appear as usual; any misfit would show as a fringe pattern between the object itself and the image from the hologram. If the blade is now set in motion and the blade illuminated by a stroboscope at the moment it is in line with the holographic image any strain in the moving blade can be measured. This amazing tool

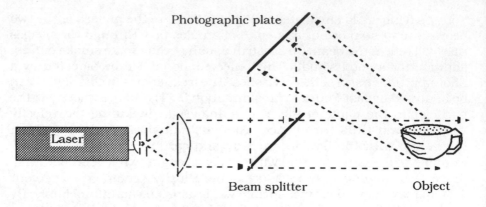

Fig. 1.28 The arrangement for taking a hologram

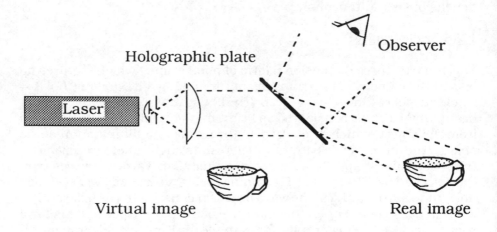

Fig. 1.29. Arrangement for viewing a hologram.

allows flow visualisation, vibration analysis and many other unique measurements to be made. Yet for some reason it is little used.

1.4.7. Speckle Interferometry

When laser light is shone onto an object the object appears speckled due to the coherent nature of the laser light forming interference patterns on the retina of the eye. These patterns are a function of the roughness of the object being viewed. If this pattern is stored in a computer and played back on a video screen at the same time as the image is received a millisecond or so later, then any movement of the object during that time will be recorded as a fringe pattern. This is a tool similar to

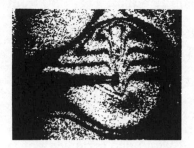

Fig. 1.30. A speckle interferogram of an electric lamp.

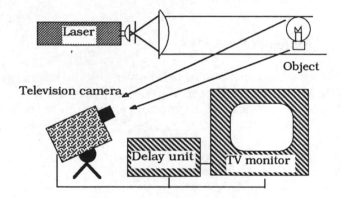

Fig. 1.31. Arrangement for speckle interferometry.

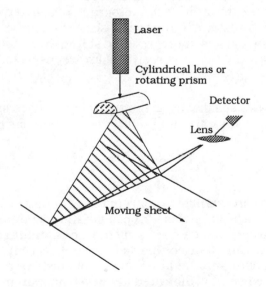

Fig. 1.32. A line image or scan inspection system.

holographic interferometry but does not require a film. Fig. 1.30 from Professor Butters' original work (11) shows the flow inside and outside a lit electric light bulb as visualised using speckle interferometry. The experimental arrangement is shown in Fig. 1.31.

1.4.8. Inspection

A laser beam is either focussed as a line or scanned over the surface of an object, for example a foil strip from a roll; any flaw will show by a variation in the reflectivity of the surface. A simplified arrangement is shown in Fig. 1.32. Scanning systems are now common in supermarket check outs for reading bar codes.

1.4.9. Analytic Technique

A focussed pulsed beam of reasonable power but little energy (approximately 10kW) strikes the material to be analysed and evaporates a small part of the surface. This vapour is sucked into a mass spectrometer and analysed.

An alternative technique is to cause a laser generated spark on the surface of the material and analyse the spectrum.

Another aspect of analysis is the analysis of fingerprints for forensic work. Fingerprints are essentially sets of curved lines the Fourier transform of these lines will be a series of dots as with an optical grating. Thus passing a laser beam through a photograph of a fingerprint would produce the Fourier transform and hence a series of dots set at various angles. This form of data is more readily stored in a computer for comparison purposes. TV colour screen masks are inspected using the same principle.

Another form of analysis is discussed under medical uses of lasers for cell cytometry (Section 1.4.13); and pollution detection has been noted in Section 1.4.4.

1.4.10. Recording

The CD player is now an established part of many homes. The compact disc (CD) is drilled with tiny holes - approximately 0.6μm diameter - using a focussed laser. The holes are drilled into tantalum sheet. The spacing is around 1.6μm. (For comparison an old 78 rpm record had tracks spaced at 150μm and the later 33 rpm records were spaced at 30μm.) The perforated sheet is mounted inside a transparent plastic for protection. The reading of the disc is by a diode laser focussed onto the surface of the disc and the signal "hole" or "no hole" is read by a photodetector either by reflection or transmission through the disc. The digital signal thus generated is the high fidelity signal required. The storage density of this system is substantial and may revolutionise the storing of lecture slides, archive material and raw data. The present challenge is to design a system which will both record and play back for the domestic market - a very real prospect in the next year or so.

Another form of recording data is in the use of holographic memories for computers. These memories are based on material whose refractive index changes when intense light strikes it. Lithium Niobate is such a material. If two crossed laser beams penetrate the crystal a form of three dimensional memory results. The storage capacity of this has the potential of being the heart of a wrist watch computer of significant power, but that is for the future. Photonic computing is however a subject of intense research since it promises to be faster than electronic computers, with fewer interconnection problems and it is less vulnerable to outside radiation (12).

1.4.11. Communications

The beam from a carbon dioxide laser has a frequency of 10^{13} Hz. Most of us are content to listen to a radio station operating at around 1MHz. Thus there is the potential for 10^7 1MHz wavebands all operating simultaneously on one laser beam - if only we could read every wave of a light beam separately! Nevertheless this accounts for the considerable interest in laser beams for communication. If this level of technology could be achieved then there would be an explosion in communication systems. The television telephone would be common place. Currently only a hundred or so separate signals can be operated simultaneously in one optical fibre from one laser. Low loss optical fibre technology with graded or stepped refractive index silica fibres is a new industry. Many new housing estates are being fitted with fibres as well as electric cables. The capacity is one attraction of optical communication; the other is the lack of interference from outside sources, the security of transmission and the ability to transmit through space.

1.4.12. Heat Source

This is the main subject of this book. The radiation from a 2kW CO_2 laser can be typically focussed to around 0.2mm diameter. The power density is then $(2000/\pi 0.1^2) = 0.6 \times 10^5 W/mm^2$. This is to be compared with 5-500W/mm² from an electric arc. Only the electron beam can rival this value. Thus the laser, even at quite modest power levels, offers the highest power density available to industry today. By defocussing, it also offers the lowest! This range of power intensity means the laser can evaporate any known material provided the beam can be absorbed or give it any specified thermal experience. It is one of the most flexible and easily automated industrial energy sources. Today it is used in production in cutting, welding, and surface treatment; as in heat treatment, melting, alloying, cladding, machining, microlithography, stereolithography, bending, texturing, engraving and marking. These processes are discussed in the following chapters.

1.4.13. Medical

Perhaps this is the area of laser applications which most people have heard about. It is certainly one of the most personal and spectacular. There are so many applications only some of them will be mentioned here. A fairly complete summary is given in reference (13).

Medical applications started with eye surgery. The blue argon laser was used to penetrate the vitreous humour with little absorption and to strike the retina. This allowed detached retinas to be welded back without the trauma of the previous freezing surgical processes. The operation became an outpatient job. Glaucoma can be treated by drilling the cornea using a CO_2 laser and thus relieving the pressure in the eye.

Surgical uses are now fairly common place. The laser is a sterile cutting tool which cauterises as it cuts. It is a non contact instrument and so cuts can be made nearer to tumours without fear of diseased cells spilling into the patient. The whole operation is thus less traumatic and damaging. Strawberry blemishes and tatoos can be removed.

The possibility of passing the laser beam from certain lasers e.g. Nd-YAG down an optical fibre allows surgical work to be done via an endoscope. Thus bleeding ulcers can be treated as an outpatient operation by encouraging the patient to swallow the endoscope and coagulating the bleeding ulcer using a YAG or argon laser. The breaking of gall stones and kidney stones can be done this way. The only mark on the patient may be a small hole for the insertion of the fibre system, though there are numerous natural holes also available.

Blocked arteries have been cleared by passing a fibre up the artery to the blockage and firing a laser at a metal plate attached to the end of the fibre. This heats the plate which allows the fatty deposit in the artery to be moved. Sleep apnea and snoring can be treated by reshaping the uvula in three or five 10 minute sessions under local anaesthetic.

So far the uses have been centred on replacing existing techniques and essentially using the laser as a knife or heater. A more scientific use of the laser in surgery is just being understood. Photodynamic Therapy (PDT) is a totally new form of medicine. In the treatment of some forms of cancer the patient is asked to take photofrin or ALA(amino laevolinic acid) drugs. This is a light sensitive drug and so the poor patient has to stay in the dark for a day or so after taking the drug. After that time the natural body tissues will have absorbed and excreted the drug. However the cancerous cells will not have done so. Thus when a laser of a specific wavelength is shone at the affected and now drug laden tumours, the radiation breaks the phosphoryn drug down so that it becomes a poison and kills the cancer, cell by cell. Those cases where this has been successful have recovered as though there had been no cancer. There is no surgical damage or scar. The process is still at the development stage. The problem is to find drugs which have good cell discrimination. The idea is excellent and full of promise - even for treating secondary cancers.

Laser based flow cytometry (14) is a new diagnostic tool which will soon be in clinical service. Flow cytometry is a technique in which pathologists illuminate a stream of droplets containing previously dyed cells. Pathological information is obtained from fluorescence measurements of the cells as they pass in single file through the light source. The information which can be obtained is cell size, cytoplasmic granulation, cell viability (including membrane integrity), cell membrane potential such as membrane viscosity, membrane and cell antigens, DNA content, cell cycle distributions, pH, DNA synthesis, total RNA content, total protein content and cellular activation including metabolism and enzyme activity. All this significant data is obtained using a multitude of dyes which mark the characteristic of interest. A flow cytometer consists of a light source (typically an argon laser), a sample chamber, optics, detectors, A to D converters (analogue to digital) and a computer for interpretation. Some of the latest commercial machines use air cooled Ar lasers at 488nm which are reasonably small and easy to maintain. They will almost certainly find their way into clinical laboratories to be used for scanning cells for cancer, AIDS, and other diseases. The latest research tools include three lasers at different frequencies (two Ar at 488nm and 565-650nm and an excimer laser in the ultra violet) and a multitude of dyes all operating simultaneously. Such a tool is capable of inspecting 100,000 cells/min and needing only 1000 molecules/cell for a signal. It can detect rare cells with specific properties in a heterogeneous population and is capable of multiple analysis. It is clearly destined to have an impact on medical pathology. In addition to this, cells can be trapped and moved by the photon pressure from low powered beams for detailed inspection or micro surgery on cell structures, such as chromosomes (15).

1.4.14.   Printing

Laser printers are in many offices. They operate on the same principle as a Xerox machine by forming an electrostatic field on a selenium drum, but with the capability of scanning 21,000 lines/min from digital data. The pattern of the field is transmitted by a rastered laser beam driven from the computer data string. The quality from these machines is as for standard print, in fact the original of this book was done this way. Desk top publishing has thus become the latest reality.

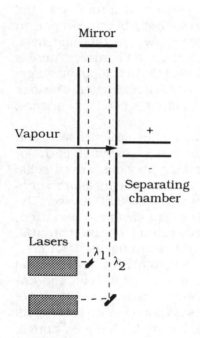

A similar rastering system can be used to make pits in photogravure rolls. This makes colour printing quick and simple and has led to a revolution in printing techniques for newspapers and other magazines.

Laser engraving of rubber rolls is now a standard technique for converting flat drawings to cylindrical rolls for the printing of wallpaper.

Fig. 1.33. Arrangement for isoptope separation by laser selective ionisation.

1.4.15. Isotope Separation

The energy levels for ionising an isotope, of say tritium or uranium required for atomic power uses, differ slightly from isotope to isotope.   By passing the isotope vapour from one chamber to another at near vacuum and shining two frequencies of laser light from carefully tuned dye lasers through the passing gas only one isotope will be ionised - not totally but with some level of efficiency. The ionised species is easily sorted by arranging an electric field in the second chamber as illustrated in the Fig. 1.33.

1.4.16. Atomic Fusion

The large AURORA excimer laser, the SHIVA CO_2 laser and NOVA glass lasers at the Lawrence Livermore and Los Alamos laboratories in the USA were built with the hope of being able to squash deuterium to produce helium with the release of the mass difference as energy according to Einstein's law of $E=mc^2$. The massive NOVA laser oscillator-amplifier system, the largest in the world for fusion work, runs many beams into the target chamber in which sits a deuterium pellet conveniently mounted. The beams are so timed that they all strike simultaneously. At 23-40TW of power some indication of success has been achieved (TW = tera Watts = 10^{12} W). The laser is currently used to generate X-rays which actually perform the implosion reaction - similar to the reaction in a hydrogen bomb. Ultimately the idea is to drop the pellets into the beam to have a maintainable power supply. There is still a long way to go!

1.5. Market for Laser Applications

The overall sales of lasers into these many application areas is roughly shown in Fig. 1.34. The division of applications within material processing is illustrated in Fig. 1.35. The market for lasers in material processing has been a growth area for several years and is expected to

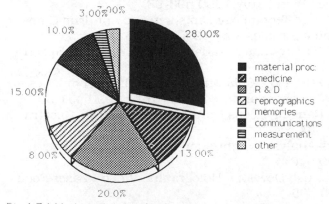

Fig.1.34. Market division for laser applications, 1987(16).

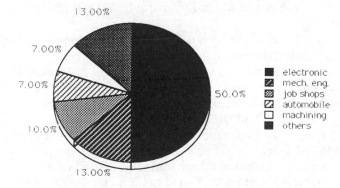

Fig.1.35. Main fields of present applications in material processing in Europe(16).

continue at 10-20%/yr for some time yet.

It is hoped that this brisk summary of the uses to which lasers have been put will let the reader see the immense new area of applications opened up by the laser and set in perspective the material processing applications discussed in the rest of this book. It may also stimulate the imagination to see some new potential applications and ideas for material processing. One thing is certain and that is that this subject is still in its infancy.

References

1. Kogelnik.H., Li.T. "Laser Beams and Resonators" App Phys $\underline{5}$ 10 p1550 Oct 1966.
2. Einstein.A. Phys Z. $\underline{18}$ 121 1917.
3. Dobbins.W.P."The only Commercial Slab Laser on the Market" Industrial Laser Review July 1990 p13-14.
4. Steinmetz.C.R. "Laser Interferometry Operates at Submicrometer Level" Laser Focus World July 1990 p93-98.
5. Morrison.D.C. "Laser Technology Enlists in the Antidrug Campaign" Lasers and Optronics May 1990 pp31-32.
6. Waggoner.J. "SDIO Says Laser Radar Works" Photonics Spectra July 1990 p18.
7. Nordstrom.R.J., Berg.L.J. "Coherent Laser Radar: Techniques and Applications" Lasers and Optronics June 1990 p51-56.
8. Arndt.W. "Laser Ranging Keeps Cars Apart" Photonics Spectra July 1990 p133-134.
9. Dance.B. "Lasers Analyse Combustion Products" Laser Focus World, July 1990 p40.
10. Carts.Y.A. "Media Lab Develops Holographic Video" Laser Focus World May 1990 p95.
11. Butters.J.N. "Speckle Interferometry and Other Technologies" Proc 1st Int Conf on Lasers in Manufacturing (LIM1) Brighton Nov 1983 pp149-160.
12. Troy.C.T. "Pentagon Calls Photonics a Critical Technology" Photonics Spectra July 1990 p83.
13. Petty.H.R., Krevsky.B. "Survey of Laser Applications in Biomedicine" Lasers and Optronics April 1990 p63-79.
14. Speser.P. "Laser based Flow Cytometry Advances Diagnostic Pathology" Laser Focus World July 1990 p.21.
15. Lewis.R. "With Lasers You Can Operate on Cells" Photonics Spectra July 1990 p74-78.
16. Steen.W.M. "Summary Review of Laser Material Processing in Europe" Proc ICALEO'89, Orlando, Florida, USA Oct 1989. publ LIA Florida USA p2-13.

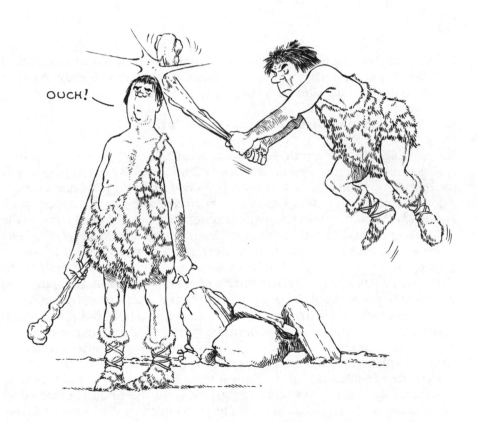

"Stimulated emission is not a new phenomenon".

Chapter 2

Basic Laser Optics

"Open the second shutter so that more light can come in"
Attributed as the dying words of Johann Wolfgang Von Goethe 1749-1832

In this chapter the basic principles of light generation and absorption are described and the fundamentals of how such energy can be manipulated in direction and shape are presented.

2.1. The Nature of Electromagnetic Radiation

Electromagnetic radiation has been a puzzle ever since man first realised it was there. Fermat (1608-1665) stated the principles of ray propagation *"The path taken by a light ray in going from one point to another through any set of media is such as to render its optical path equal, in the first approximation, to other paths closely adjacent to the actual one"*. This is a rather complicated statement from which the laws of reflection and refraction can be derived. Huygens (1690) (1) introduced the wave concept of light to explain refraction and reflection. Newton in 1704 unravelled the puzzle of colour (2). Einstein (3) in 1905 invented the concept of the photon to explain the photoelectric effect. In fact there is still some mystery left. For example if light passes through a slit and then falls on a screen, as in Young's famous experiment, a diffraction pattern is formed on the screen. The phenomenon can be simply explained by assuming that the radiation falling on the slit is a waveform of radiation. It is difficult to explain the experiment by assuming the light is a stream of particles. In the photoelectric effect light falling on a target will give off electrons of energy E from the target regardless of the intensity, where E is given by:

$$E = h\nu - p \qquad (2.1.)$$

where: h = Planck's constant 6.625 x 10^{-34} J.s,
           ν = frequency = c/λ,
           p = constant characteristic of the material,
           c = velocity of light, 2.99 x 10^8 m/s,
           λ = wavelength of light, m.

In the wave theory, the radiation would be spread over the surface and would not all be available for one electron.

This dichotomy between waves and particles varies in significance with the wavelength or energy of the "photons". Thus at the long wavelengths from radio to blue light, the wave theory explains most phenomena observed for normal intensities. With X-rays and gamma rays, which have highly energetic photons of short wavelength, the particle theory explains most events.

The quantum theory, of which we are talking here, was initiated by Heisenberg and Schrodinger (4) in 1926. It makes a link between these states through Bohr's analysis of Planck's constant. He suggested that the constant is the product of two variables, one characteristic of the wave and the other of a particle. Thus, if the wave has a period, T, a wavelength, λ, particle energy, E, and momentum , p, Bohr suggested, on dimensional grounds amongst others, that h = ET = $p\lambda$. Thus if the particle aspects are strong, then the wave aspects will be weak. It just happens that the size of Planck's constant is such that the electromagnetic spectrum takes us from strongly particle type radiation to strongly wave type radiation. Why Planck's constant is of such a size is unknown and must be left as an exercise for the reader and his heirs and successors! However this concept that λ = h/p suggests all matter with momentum has a wavelength. This was shown to be the case for electrons by Davisson and Germer in the USA and G.P.Thomson in the UK, but the size of "h" makes the wavelength very small. The wavelength of the earth for example would be:

Mass of the earth	m =	5.976 x 10^{24}	kg
Velocity of earth	v =	3 x 10^4	m/s

∴   Earth's wavelength, λ = 6.625 x 10^{-34}/(5.976 x 10^{24} x 3 x 10^4)
                  = 3.7 x 10^{-63} m

A bit difficult to measure!

The momentum of a photon can be found from Planck's law E = hν (justified from the photoelectric effect and other phenomena, where ν = frequency) and Einstein's equivalence of mass and energy E = mc^2

(justified by the experiments on nuclear disintegration).

Together these give: $h\nu = h\,c/\lambda = mc^2$ (2.2.)

and since the momentum , $p = mc$, we have $p = h/\lambda$ - the same as Bohr's relationship quoted above.

Incidentally this suggests that the pressure, P, on a mirror from a photon from a carbon dioxide laser incident normally is (see Section 7.2.1.2.) $2p = P = 2 \times 6.625 \times 10^{-34}/10.6 \times 10^{-6} = 1.25 \times 10^{-28}$ Ns/photon, not of any great significance until one considers the avalanche of photons possible with the laser.

The energy of a photon from a carbon dioxide laser is calculated in Table 2.1 to be 1.85×10^{-20} J, where it is compared with photons from other optical generators. Thus in a 1kW CO_2 laser beam there will be a flux of $1000/1.85 \times 10^{-20} = 5 \times 10^{22}$ photons/s and the overall force will be 6×10^{-6} Pa. Still not very exciting but possibly measurable. Over the focussed spot from this laser, of say 0.1mm diameter, the pressure would be: $(4 \times 6 \times 10^{-6})/(p[0.1 \times 10^{-3}]^2) = 760$N/m². This is equivalent to a depression in molten steel of approximately 1cm! Very close to what is observed. One wonders whether we have missed something in ignoring photon pressure.

Table 2.1		Photon Properties of Different Lasers			
	Source	Wavelength	Frequency		Energy
Device	of Laser Energy	λ	ν		Ep*
		micron	Hz	eV	J x 1.0 E20
Cyclotron	Accelerator	0.1(X-ray)	2.9 E15	12.3	192
Free Electron Laser (FEL)	Magnetic wiggler	1 E3-6	1 E8-11	1 E-6	1 E-2 to -5
Excimer		0.249 (u/v)	1.2 E15	4.9	79.4
Argon	Atomic electron orbits	0.488 (blue)	6.1 E14	2.53	40.4
He/Ne		0.6328 (red)	4.7 E14	1.95	31.1
Nd-YAG	Molecular vibration	1.06 (IR)	2.8 E14	1.16	18.5
CO		5.4	5.5 E13	0.23	3.64
CO2		10.6	2.8 E13	0.12	1.85
* Energy calculated from Ep = hν; 1eV = 1.6 E-19 J					

It is assumed that the velocity of a photon is always c, the velocity of light in a vacuum and that this is a universal constant. Photons do not behave as normal particles which can have a variable velocity. The early explanations of refraction, for example, in which the wave theory explains the process by suggesting that the velocity of light varies from one medium to another has to be interpreted as: the photon travels at the speed, c, always; but in passing through a medium the wave front slows due to the absorption/re-emission processes taking place as the photon interacts with the molecules of the medium through which it travels. The reason for this universal constant is related to the concept of time; it has an uncanny ring that we have more thinking to do to understand this subject.

2.2. Interaction of Electromagnetic Radiation with Matter

When electromagnetic radiation strikes a surface the wave travels as shown in the Fig. 2.1. Some radiation is reflected, some absorbed and some transmitted. As it passes through the new medium it will be absorbed according to some law such as Beer Lambert's law, $I = I_o e^{-\beta z}$. The absorption coefficient, β, depends on the medium, wavelength of the radiation and the intensity. The intensity can be significant with high energy density beams as in laser processing. At higher intensities multi photon interactions are more likely to occur, causing non linear optical events such as Rayleigh scattering, Brillouin scattering and Raman scattering. The manner in which this radiation is absorbed is considered to be as follows. Electromagnetic radiation can be represented as an

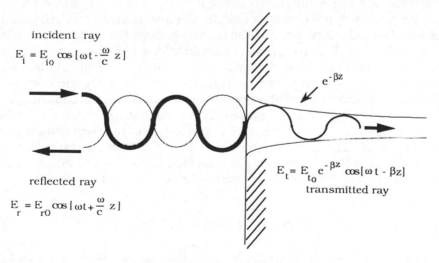

Fig. 2.1. The phase and amplitude of an electromagnetic ray striking an air/solid interface and undergoing reflection and transmission.

electric vector field and a magnetic vector field illustrated in Fig. 2.2. When it passes over a small elastically bound charged particle, the particle will be set in motion by the electric force from the electric field, E. Provided that the frequency of the radiation does not correspond to a natural resonance frequency of the particle then fluorescence or absorption will not occur but a forced vibration would be initiated. The force induced by the electric field, E, is very small and is incapable of vibrating an atomic nucleus. We are therefore discussing photons interacting with electrons either free or bound. This process of photons being absorbed by electrons is known as the "Inverse Bremsstrahlung Effect" (The Bremsstrahlung Effect is the emission of photons from excited electrons.) As the electron vibrates so it will either reradiate in all directions or be restrained by the lattice phonons (the bonding energy within a solid or liquid structure). In this latter case the phonons will cause the structure to vibrate and this vibration will be transmitted through the structure by the normal diffusion type processes due to the linking of the molecules of the structure. The vibrations in the structure we detect as heat.

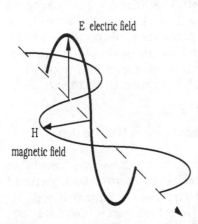

E electric field

H
magnetic field

Fig. 2.2. The electric and magnetic field vectors of electromagnetic radiation.

The flow of heat is described by Fourier's laws on heat conduction - a flux equation ($q/A = -kdT/dx$). If sufficient energy is absorbed then the vibration becomes so intense that the molecular bonding is stretched so far that it is no longer capable of exhibiting mechanical strength and the material is said to have melted. On further heating the bonding is further loosened due to the strong molecular vibrations and the material is said to have evaporated. The vapour is still capable of absorbing but only slightly since it will only have bound electrons; the exception occurs if the gas is sufficiently hot that electrons are shaken free and the gas is then said to be a plasma. Plasmas can be strongly absorbing if their free electron density is high enough. The electron density in a plasma is given by equations such as the Saha equation (5), which assumes thermal equilibrium in the plasma so that standard free energy changes can be calculated using conventional thermodynamic principles.

$$\ln\left(\frac{N_i}{N_o}\right)^2 = -5040\left[\frac{V_i}{T}\right] + 1.5\ln(T + 15.385) \qquad (2.3.)$$

where N_i = ionisation density; N_o = density of atoms; V_i = ionisation potential, eV; T = absolute temperature, K. This indicates that tempera-

tures of the order of 10,000-30,000°C
are required for significant absorption (Fig. 2.3) (6). This sequence in
the stages of absorption is illustrated in Fig. 2.4.

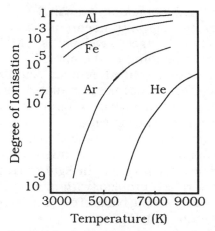

Fig. 2.3. Degree of ionisation as a
function of temperature.

It is interesting to note that the
energy absorbed by an electron may
be that of one or more photons.
However, it will only be in extreme
cases, such as the NOVA laser operating at 23-40 TW or so (T stands for
tera = 10^{12}) that a sufficient numbers of photons would be simultaneously absorbed to allow the emission of X-rays during laser processing. This is a strategic advantage for
the laser over Electron Beam processes which require shielding against
this hazard.

Incidentally the mean free time of
electrons in a conductor is calculated to be around 10^{-13} s. This means
that only for extremely short laser
pulses of around 1ps (pico second,
10^{-12} s per pulse) is it possible that
the material would contain two temperatures not at equilibrium - the
electron temperature and the atomic
temperature. Also for very short
pulses non Fourier conduction has
been postulated (7) in which a compression or heat wave forms; this
may be related to the acoustic signals
noted in Section 7.2.1.2 or shock
hardening mentioned in Section 6.1.

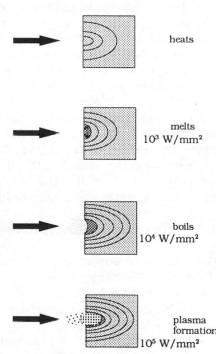

Fig. 2.4. Sequence of absorption events
varying with absorbed power.

2.2.1. Non Linear Effects

2.2.1.1 Fluorescence: Certain materials will absorb specifically at the
frequency of the incident beam. The structure thus becomes excited. To
lose this energy some materials may reradiate at a different frequency.
This effect is known as fluorescence. One application is in dye lasers.

2.2.1.2. Stimulated Raman Scattering: If low intensity light is transmitted through a transparent material, a small fraction is converted into light at longer wavelengths with the frequency shifted (Stokes shift) corresponding to the optical phonon frequency in the material. This process is called Raman scattering. At higher intensities, Raman scattering becomes stimulated, and from the spontaneous scattering a new light beam can be built up. Under favourable conditions, the new beam can become more intense than the remaining original beam. The amplification is equally high in the forward or the backward directions. This may lead to a situation where a large fraction of the radiation is redirected towards the light source rather than towards the target. This could be a problem with intense light being transmitted in fibres, but also forms the basis of certain detection techniques, such as LIDAR (see Section 1.4.4).

2.2.1.3. Stimulated Brillouin Scattering: The same process takes place with the acoustical phonons. The corresponding frequency shift is much smaller. Acoustical phonons are sound waves, and the frequency shift exists only for the wave in the backward direction. Again at high intensities the Brillouin effect becomes a stimulated process, and the Brillouin wave may get much more intense than the original beam. Almost the entire beam may be reflected towards the laser source.

2.2.1.4. Rayleigh Scattering: Particles smaller than the wavelength of the incident light will scatter the radiation in the form of a spherical wave. The extent of this power loss depends on the number of the particles and the wavelength. It has been found that this effect is proportional to $1/\lambda^4$. It is the reason the sky is blue, but it can also be a limiting factor in the design of fibres and some optics.

2.3. Reflection (or Absorption)

The value of the absorption coefficient will vary with the same effects that affect the reflectivity.
For opaque materials the reflectivity = 1 - absorptivity and for transparent materials, reflectivity = 1- (transmissivity + absorptivity).

In metals the radiation is predominantly absorbed by free electrons in an "electron gas". These free electrons are free to oscillate and reradiate without disturbing the solid atomic structure. Thus the reflectivity of metals is very high in the waveband from visible to the DC, see Fig. 2.5. As a wavefront arrives at a surface then all the free electrons in the surface vibrate in phase generating an electric field 180° out of phase with the incoming beam. The sum of this field will be a beam whose angle of reflection equals the angle of incidence. This "electron gas" within the

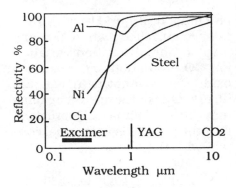

Fig. 2.5. Reflectivity of a number of metals as a function of wavelength

metal structure means that the radiation is unable to penetrate metals to any significant depth, only one to two atomic diameters. Metals are thus opaque and they appear shiny.

The reflection coefficient for normal angles of incidence from a dielectric or metal surface in air (n = 1) may be calculated from the refractive index, n, and the extinction coefficient, k (or absorption coefficient as described above) for that material:

$$R = \{(1-n)^2 + k^2\}/\{(1+n)^2 + k^2\} \tag{2.4.}$$

For an opaque material such as a metal the absorptivity, A is:

$$A = 1 - R$$

$$A = 4n/\{(n+1)^2 + k^2\} \tag{2.5.}$$

Some values of these constants are given in Table 2.2. The value of the reflectivity, R, is for a perfectly flat clean surface - which is rarely the case.

2.3.1. Effect of Wavelength

At shorter wavelengths, the more energetic photons can be absorbed by a greater number of bound electrons and so the reflectivity falls at shorter wavelengths and the absorptivity of the surface is increased (Fig. 2.5.).

2.3.2. Effect of Temperature

As the temperature of the structure rises there will be an increase in the phonon population causing more phonon-electron energy exchanges. Thus the electrons are more likely to interact with the structure rather than oscillate and reradiate. There is thus a fall in the reflectivity and an increase in the absorptivity with a rise in temperature, as seen in Fig. 2.6 (9).

2.3.3. Effect of Surface Films

The reflectivity is essentially a surface phenomenon and so surface films may have a large effect. Fig. 2.7 shows that for interference coupling the

Table 2.2.

Complex refractive index and reflection coefficient for some materials to 1.06μm radiation (8).

Material	k	n	R
Al	8.50	1.75	0.91
Cu	6.93	0.15	0.99
Fe	4.44	3.81	0.64
Mo	3.55	3.83	0.57
Ni	5.26	2.62	0.74
Pb	5.40	1.41	0.84
Sn	1.60	4.70	0.46
Ti	4.0	3.8	0.63
W	3.52	3.04	0.58
Zn	3.48	2.88	0.58
Glass	0	1.5	0.04

film must be around $[(2n + 1)/4]\lambda$ to have any effect, where n is any integer. The absorption variation for CO_2 radiation by a surface oxide film is shown in Fig. 2.8 (9,10). One of these surface films may be a plasma (11) provided that the plasma is in thermal contact with the surface.

2.3.4. Effect of Angle of Incidence

The variation of the reflectivity with angle of incidence is shown in Fig. 2.9. At certain angles the surface electrons may be constrained from vibrating since to do so would involve leaving the surface. This they would be unable to do without disturbing the matrix, i.e. absorbing the photon. Thus, if the electric vector is in the plane of incidence, the vibration of the electron is inclined to interfere with the surface and absorption is thus high; while, if the plane is at right angles to the plane of incidence then the vibration can proceed without reference to the surface and reflection is prefered. There is a particular angle - the **"Brewster"** angle - at which the angle of reflection is at right angles to the angle of refraction. When this occurs it is impossible for the electric vector in the plane of incidence to be reflected since there is no component at right angles to itself. Thus the reflected ray will have an electric vector only in the plane at right angles to the plane of incidence. At this angle the angle of refraction = (90° - angle of incidence) and hence by Snell's law (see Section 2.4.) the refractive index, n = tan(Brewster angle). Any beam which has only or principally one plane for the electric vector is called a **"polarised"** beam. Some values of the refractive index and the Brewster angles for different materials is given in Table 2.3.

Most lasers produce beams which are polarised due to the nature of the amplifying process within the cavity which will favour one plane. Any plane will be favoured in a random manner, unless the cavity has folding mirrors, in which case the electric vector which is at right angles to the plane of incidence on the folded mirrors will be favoured because that is the one suffering the least loss.

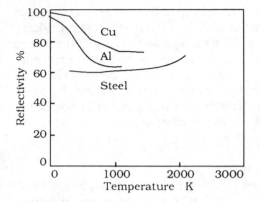

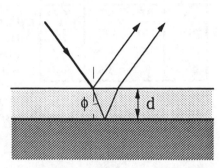

Fig. 2.6. Reflectivity as a function of temperature for 1.06μm radiation.

Fig. 2.7. A surface film acting as an interference coupling, "anti - reflection" coating. If $2d/\cos\phi = [(2n + 1)/2]\lambda$ then there will be destructive interference of the reflected ray.

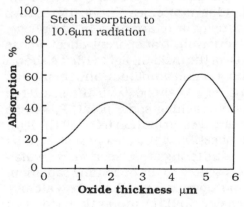

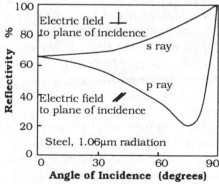

Fig. 2.8. Absorption as a function of the thickness of an oxide film on steel for 1.06μm radiation.

Fig. 2.9. Reflectivity of steel to polarised 1.06μm radiation.

Table 2.3.	Refractive Index and Brewster Angles for Various Materials	
Material	Refractive Index (1μm radiation)	Brewster Angle
	k n	
Al	8.5 1.75	60.2
Fe	4.49 3.81	75.2
Ti	3.48 2.88	70.8
Glass	- 1.5	56.3

2.3.5. Effect of Materials and Surface Roughness

As noted in Sections 2.3.2 and 6.2 roughness has a large effect on absorption due to the multiple reflections in the undulations. There may also be some "stimulated absorption" due to beam interference with sideways reflected beams(12). Provided the roughness is less than the beam wavelength, the radiation will not suffer these events and hence will perceive the surface as flat.

2.4. Refraction

On transmission the ray undergoes refraction described by Snell's law: *"The refracted ray lies in the plane of incidence, and the sine of the angle of refraction bears a constant ratio to the sine of the angle of incidence".*

$$\sin\phi/\sin\varphi = n = v_1/v_2 \tag{2.6.}$$

where: n = Refractive index.
           ϕ = Angle of incidence.
           φ = Angle of refraction.
           v_1 = Apparent speed of propagation in medium 1.
           v_2 = Apparent speed of propagation in medium 2.

The apparent change in velocity of light as it passes through a medium is the result of scattering by the individual molecules. The scattered rays interfere with the primary beam causing a retardation in the phase. Consider a plane wave striking a very thin, transparent sheet whose thickness is less than the wavelength of the incident light (13), as shown in Fig. 2.10. Let the electric vector have a unit amplitude and then it can be represented at a particular time as $E = \sin2\pi x/\lambda$. If the scattered intensity is small, then the intensity reaching some point, P, will be essentially the original wave plus a small contribution from all the light scattered from all the atoms of the sheet. Now the energy scattered by one atom will be proportional to its scattering cross section, σ, which is that part of the area of the atom presented to the oncoming radiation. Thus the scattered amplitude is proportional to $\sqrt{\sigma}$. If there are N atoms/cm³, the total scattered amplitude per cm² would be proportional to $\sqrt{\sigma}Nt$; where t is the thickness. Since it is assumed that $t \ll \lambda$ the wave leaving the sheet will all be in phase. At P, however their phases will differ by the different distances travelled, R. We can calculate the net effect by summing the scattered amplitudes of all the atoms over the surface, E_s

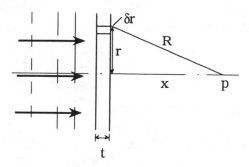

Fig. 2.10. Radiation passing through a thin transparent layer.

- allowing for the amplitude being proportional to $1/R$:

$$E + E_s = \sin \frac{2\pi x}{\lambda} + \sqrt{\sigma}\ Nt \int_0^\infty \frac{2\pi r dr}{R} \sin \frac{2\pi R}{\lambda}$$

since $x^2 + r^2 = R^2$, and x is constant, we have $rdr = RdR$, and the integral may be written:

$$\int_0^\infty \frac{2\pi}{R} \sin \frac{2\pi R}{\lambda} rdr = 2\pi \int_x^\infty \sin \frac{2\pi R}{\lambda} dR = \frac{2\pi\lambda}{2\pi} \left[-\cos \frac{2\pi R}{\lambda} \right]_{R=x}^{R=\infty}$$

(The integral limits are 0 to ∞ for r and x to ∞ for R).

At $R = \infty$, the quantity in brackets is equal to zero and so we have:

$$E + E_s = \sin \frac{2\pi x}{\lambda} + \sqrt{\sigma}\ Nt\lambda \cos \frac{2\pi x}{\lambda}$$

This is of the form: $\sin A + B \cos A$, where B is assumed very small. Under these conditions we may write:

$$\sin(A+B) = \sin A \cos B + \cos A \sin B \approx \sin A + B \cos A$$

Therefore:
$$E + E_s = \sin \left(\frac{2\pi x}{\lambda} + \sqrt{\sigma}\ Nt\lambda \right)$$

which shows that the phase of the wave at P has been altered by the amount $\sqrt{\sigma}Nt\lambda$. But we know that the presence of a sheet of refractive index, n, and thickness, t, would have retarded the phase by:
$$2\pi(n-1)t/\lambda$$

hence: $\sqrt{\sigma}\ Nt\lambda = \dfrac{2\pi}{\lambda}\,(n - 1)\,t$

and so: $n - 1 = \dfrac{1}{2\pi}\,N\lambda^2\,\sqrt{\sigma}$ (2.7.)

This derivation is not precise (it has not allowed for absorption) but it has shown the nature of the refraction process and how the material properties affect the refractive index. For example introduce a strain and the value of N may vary and so on. It does not show how n varies with λ since the scattered intensity does not just depend upon σ but also on $1/\lambda^4$ - Rayliegh Scattering Law. The normal form of a dispersion curve (refractive index vs wavelength) is known as a Cauchy Equation:

$$n = A + B/\lambda^2 + C/\lambda^4.$$

2.5. Laser Beam Characteristics

The energy from a laser is in the form of a beam of electromagnetic radiation. Apart from power it has the properties of wavelength, coherence, mode/diameter and polarisation. These are now discussed in the following sections.

2.5.1. Wavelength

Since the invention of the laser in 1960 there are many hundreds of lasing systems but only a few of commercial significance in material processing. Some of the wavelengths of the important material processing lasers are shown in Fig.1 of the Prologue and the levels of power achieved to date are illustrated in Fig. 1.19.a.

The wavelength depends on the transitions taking place by stimulated emission. The wavelength may be broadened by Doppler effects or by related transitions from higher quantised states as with the CO laser. On the whole the radiation from a laser is amongst the purist spectral forms of radiation available. Very high spectral purity can be achieved by using a frequency selecting grating as the rear mirror, but this is never, if ever, worth the effort for material processing. In consequence if one wishes to achieve a very short pulse of light, for example of a femto second (10^{-15} s, a beam of light around 0.3μm long!), it is not possible without first making a laser with a broader waveband as is required by the Fourier series which defines such a short pulse waveform. But that is a problem for others who are not so involved in material processing.

2.5.2. Coherence

The stimulated emission phenomenon means that the radiation is
generating itself and in consequence
a continuous waveform is possible
with low order mode beams. The
length of the continuous wavetrain
may be many metres long. The comparison of laser light to standard
random light is illustrated in Fig.
2.11. This long coherence length allows some extraordinary interference
effects with laser light as noted in
Chapter 1, such as length gauging,

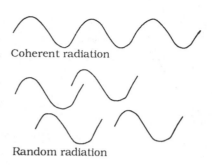

Coherent radiation

Random radiation

Fig. 2.11. Comparison of the electric
vector phase for coherent and random
radiation.

speckle interferometry, holography and Doppler velocity measurement.
This property has not yet been used in material processing. In years to
come it may be that someone will be able to use it as a penetration meter,
or to do subtle experiments with interference banded heat sources.

2.5.3. Mode and Beam Diameter

A laser cavity is an optical oscillator. When it is oscillating there will be
standing electromagnetic waves set up within the cavity and defined by
the cavity geometry. It is possible to calculate the wave pattern for such
a situation and it is found that there is not only a longitudinal standing
wave but also a transverse one. For a non amplifying, cylindrical cavity
the amplitude of the transverse standing wave pattern, $E(r,\phi)$, is given
by a Laguerre-Gaussian Distribution Function of the form:

$$E(r,\phi) = E_o \left(\frac{\sqrt{2}\,r}{w(z)} \right)^n L_p^n \left(\frac{2r^2}{w^2(z)} \right) \exp \left(-\frac{r^2}{w^2(z)} \right) \left(\left\{ {\sin \atop \cos} \right\} n\phi \right) \qquad (2.8.)$$

where:

$E(r,\phi)$	=	Amplitude at point r,ϕ
$w(z)$	=	Beam diameter at point z along beam path
r	=	Radial position
n	=	An integer
ϕ	=	Angular position

where:
$$L_p^n (x) = e^x \frac{x^{-n}}{p!} \frac{d^p}{dx^p} (e^{-x} x^{p.n})$$

which is the generalised Laguerre Polynomial. Some low order polynomials are:

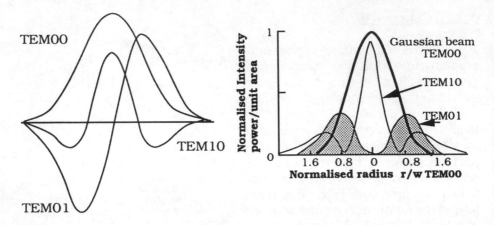

Fig. 2.12. Amplitude variation for various modes.

Fig. 2.13. Intensity distribution for various modes.

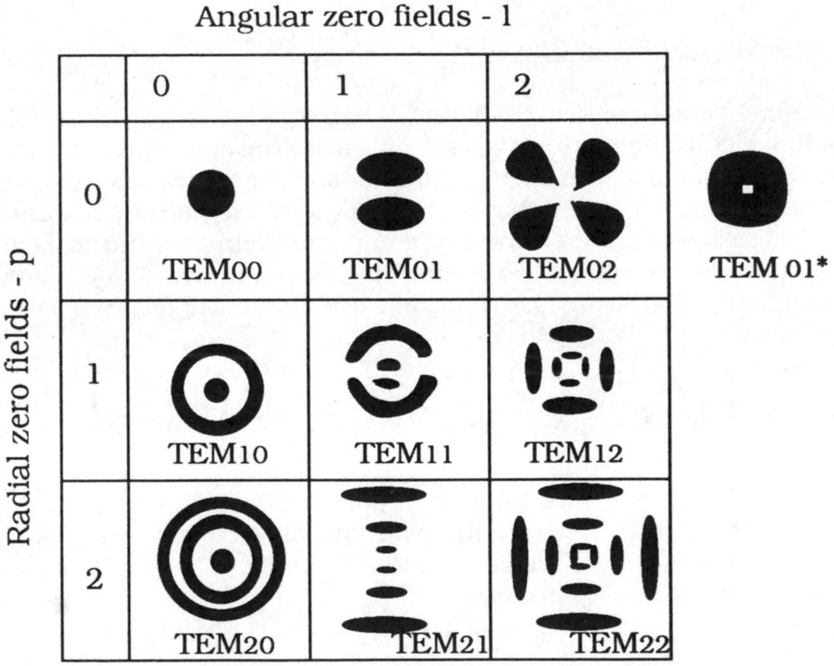

Angular zero fields - l

Radial zero fields - p	0	1	2	
0	TEM00	TEM01	TEM02	TEM 01*
1	TEM10	TEM11	TEM12	
2	TEM20	TEM21	TEM22	

Fig. 2.14. Various mode patterns.

$L_0^n(x) = 1$
$L_1^n(x) = n + 1 - x$
$L_2^n(x) = 1/2(n + 1)(n + 2) - (n + 2)x + 1/2(x^2)$

The intensity distribution is found from the square of the amplitude:

$$P(r,\phi) = E^2(r,\phi)$$

These are the classical mode distributions for a circular beam. (A square beam is similar but with Hermite Polynomials.) A plot of the amplitude and spatial intensity distributions which this expression represents for various orders of mode is shown in Figs. 2.12 and 2.13. Typical mode patterns that would be made from such beams are shown in Fig. 2.14.

The classification of these Transverse Electromagnetic Mode patterns (TEM_{plq}) is:

p = Number of radial zero fields
l = Number of angular zero fields
q = Number of longitudinal zero fields

Most slow flow (SF) lasers operate with a near perfect TEM00 or TEM01* mode. The TEM01* mode is made from an oscillation between two orthogonal TEM01 modes as illustrated in Fig. 2.14.

Most fast axial flow (FAF) lasers also give a beam with a low order mode since they have long, narrow tubes - low Fresnel number $(a^2/\lambda L)$ - see Section 1.1.1. The modes from these lasers may be slightly distorted due to plasma density variations.

Transverse flow (TF) lasers usually have multimode beams of indeterminate ranking. They are either quasi-Gaussian - in that they are a single lump of power - usually asymmetric due to the transverse amplification being different from side to side because of heating or they may be ring shaped if the cavity is engineered to be an unstable cavity - see Section 1.1.1.

The higher the order of the mode the more difficult it is to focus the beam to a fine spot, since the beam is no longer coming from a virtual point.

A question arises in material processing as to what is the beam diameter. For example Figs. 2.12 and 2.13 were calculated with the mathematical diameter, w(z), the same. This is obviously not related to the diameter which affects heating processes. Sharp (14) argues that the beam diameter should be defined as that distance within which $1/e^2$ of the total power exists.

2.5.4. Polarisation

The stimulated emission phenomenon not only produces long trains of waves but these waves will also have their electric vectors all lined up. The beam is thus polarised. Many of the early lasers and some of the more modern which do not have a fold in the cavity will produce randomly polarised beams. In this case the plane of polarisation of the beam changes with time - and the cut quality may show it! To avoid this it is necessary to introduce into the cavity a fold mirror of some form. Outside the cavity such a fold would make no noticeable difference. Inside the cavity it is a different matter since the cavity is an amplifier and hence the least loss route is the one being amplified in preference to the other - in fact almost to its total exclusion. Polarised beams have a directional effect in certain processes for example cutting due to the reflectivity effects shown in Fig. 2.9. Hence material processing lasers are usually engineered to give a polarised beam which is then fitted with a depolariser - see later, Section 2.7.2.

2.6. Focussing with a Single Lens

In order to manipulate the beam, to guide it to the workplace and shape it, there are many devices which have so far been invented. These devices are now discussed together with the basic theory of their design. In nearly all of them the simple laws of geometric optics are sufficient to understand how they will work; but to calculate the precise spot size and depth of focus one needs to refer to Gaussian optics and diffraction theory.

2.6.1. Focal Spot Size

2.6.1.1. Diffraction Limited Spot Size: A beam of finite diameter is focussed by a lens onto a plate as shown on Fig. 2.15. The individual parts of the beam striking the lens can be imagined to be point radiators of a new wave front. The lens will draw the rays together at the focal plane and thus constructive and destructive interference will take place. When two rays arrive at the screen and they are half a wavelength out of phase then they will destructively interfere and the light intensity will fall, conversely when they arrive in phase. Thus if the ray AB (Fig. 2.15) is $\lambda/2$ longer than ray CB the point B will represent the first dark ring of what is known as a "Fraunhofer" Diffraction Pattern. The central maximum will contain approximately 86% of all the power in the beam. The diameter of this central maximum will be the focussed beam diameter, usually measured between the points where the intensity has fallen to $(1/e^2)$ of the central value.

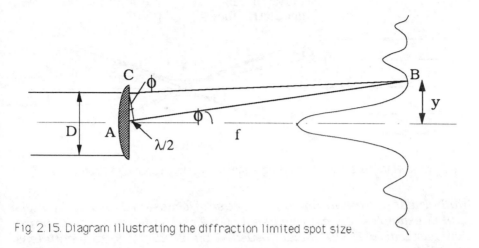

Fig. 2.15. Diagram illustrating the diffraction limited spot size.

For a rectangular beam with a plane wave front, the first dark fringe will occur when the beam path difference between the centre and the edge rays, d, is:

$$d = \lambda/2 = (D/2)\sin\phi$$

That is when: $\lambda = D \sin\phi$

or when $m\lambda = D \sin\phi$ for other fringes.

Thus: $2y = 2f \tan\phi$

and for small angles $\tan\phi = \sin\phi$

therefore $2y = d_{min} = 2f\lambda/D$

There is a correction for circular beams of 1.22 and so the equation becomes for circular beams:

$$d_{min} = 2.44\ f\lambda/D \qquad\qquad (2.9.)$$

The focal spot size for a multi mode beam will be larger because the beam is coming from a cavity having several off axis modes of vibration and therefore not all coming from an apparent point source. This correction for a TEM_{plq} beam is:

$$d_{min} = 2.44\ (f\lambda/D)\ (2p + 1 + 1) \qquad\qquad (2.10.)$$

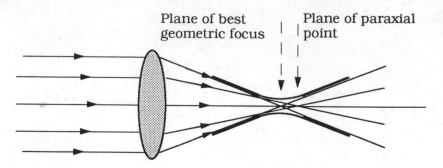

Fig. 2.16. Spherical aberration of a single lens focussing a parrallel beam.

(Radial nulls p, are more damaging to ease of focussing than angular nulls .l. For example, the expected spot size for a CO_2 laser beam 22mm in diameter with a TEM01 mode focussed by a 125mm focal length lens would be expected to be: $d_{min} = 2.44 \{(125 \times 10.6 \times 10^{-3})/22\} \times 2 = 0.29mm$ whereas a TEM10 beam would be expected to focus to 0.44mm.)

2.6.1.1.1. *M^2 concept of Beam Quality:* The fundamental propagation equation for a diffraction limited Gaussian beam (TEM00), see Fig. 2.18, is:

$$w^2(z) = w_o^2 \left[1 + \left\{ \frac{\lambda z}{\pi w_o^2} \right\}^2 \right] \tag{2.11}$$

In the far field, $z \to$ large, $\left(\frac{\lambda z}{\pi w_o^2} \right)^2 \gg 1$

Then $w(z)/z = \Theta = \lambda/\pi w_o$. In focussing a Gaussian beam $\Theta = (D_L/2)/f$. where D_L is the beam diameter on the lens. So the diffraction limited Gaussian spot size is:

$$d_{min} = 2w_o = \frac{4f\lambda}{\pi D_L} = 1.2 \, F\lambda$$

For multi mode beams the divergence, Θ_{act} is always larger than the Gaussian value, as seen from the Laguerre-Gaussian polynomials. Now since light travels in straight lines this leads to the geometric fact that $M = constant = \Theta_{act}/\Theta_{Gauss.}$
where Θ_{gauss} is the divergence of a Gaussian beam from the same cavity but not the same diameter.

$\Theta_{gauss} = \lambda/\pi w_o = \lambda/\pi(W_o/M)$ since, also by geometry $w_o = W_o/M$

Therefore the divergence of a Gaussian beam of the same size, $\Theta_r = M\Theta_{act}$ and we have:

$$M^2 = \frac{\Theta_{act}}{\Theta_r}$$

Applied to a lens we have:

$$\Theta_{act} = \frac{D_L}{2f} \text{ and } \Theta_r = \frac{2\lambda}{\pi d_{min}}$$

$$\therefore d_{min} = \frac{4M^2 f\lambda}{\pi D_L} \qquad\qquad (2.12.)$$

M can be measured from the beam diameter at two known locations along the beam. Hence this parameter, M^2, could become a useful measure of beam quality.

2.6.1.2. Spherical Aberration: There are two reasons why a lens will not focus to a theoretical point; one is the diffraction limited problem discussed above and the other is that a spherical lens is not a perfect shape. Most lenses are made with a spherical shape since this can be accurately manufactured without too much cost and the alignment of the beam is not so critical as with a perfect aspheric shape. The net result is that the outer ray entering the lens is brought to a shorter axial focal point than the rays nearer the centre of the lens, as shown in Fig. 2.16. This leaves a blur in the focal point location. The plane of best geometric focus (the minimum spot size) is a little short of the plane of the plane wavefront; the paraxial point. The size of the minimum spot, d_a is given by:

$$d_a = K(n;q;p) \left(\frac{D_L}{f}\right)^3 \quad S_s = 2\,\theta_a\,S_2 \qquad\qquad (2.13.)$$

where:
Θ_a = Angular fault (half angle).
s_2 = Distance from lens.
$K(n;q;p)$ = Factor dependent on the refractive index, n, the lens shape, q and the lens position, p.
D_L = Diameter of top hat beam mode on lens.
f = Focal length of the lens.

$$K(n;q;p) = \pm \frac{1}{128n(n-1)}\left[\frac{n+2}{n-1}q^2 + 4(n+1)pq + (3n+2)(n-1)p^2 + \frac{n^3}{n-1}\right]$$

where: q = the lens shape factor = $(r_2+r_1)/(r_2-r_1)$
$r_1,\ r_2$ = the radius of curvature of the two faces of the lens and
p = the position factor = $1 - 2f/s_2$

Fig. 2.17 shows the variation of spherical aberration with lens shape. The optimum shape is when the refraction angles at both faces of the lens are approximately equal. Note that there is a huge difference between a plano convex lens mounted one way or the opposite way around.

Other lens faults are:
1. Mechanical and optical axis are not correctly aligned.
2. Lens surface is not correctly spherical.

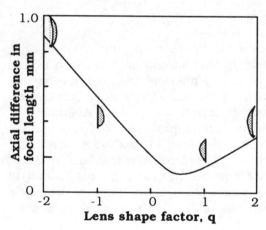

Fig 2.17. Spherical aberration of a ray 1cm off the optic axis passing through a lens of focal length 10cm, diameter 2cm and refractive index 1.517 (after 13).

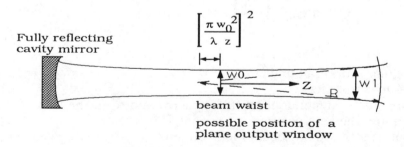

Fig. 2.18. Varying radius of curvature of the phase field with distance. A small value of R is known as the "near" field, while a large value is known as the "far" field.

2.6.1.3. Near and Far Field Effects: So far we have assumed that the beam strikes the lens with a plane wave front. If the cavity of the laser is fitted with a curved back end mirror, which is normal, then the beam will be expanding after it leaves the cavity. It will have a beam waist, w_o, as shown in Fig. 2.18. If the output window is plane then the beam waist will be situated on that mirror since it is the location at which the wave front is plane. The radius of curvature of the wavefront, R, is given by:

$$R = z[1 + (\pi w_o^2/\lambda z)^2]\qquad\qquad (2.14.)$$

Where z = distance from beam waist.

The beam radius at any location, z, is given by: $w_1 = w_o/[1 + (\lambda z/\pi w_o)^2]$ (Equation 2.11.) The effect of this wavefront curvature is to slightly alter the focal characteristics. This is known as the "near" and "far " field effects, i.e. where it is significant and where it is not.

2.6.2. Depth of Focus

The depth of focus is the distance over which the focussed beam has approximately the same intensity.   It is defined as the distance over which the focal spot size changes by ±5%.

It can be shown by geometry that this gives a distance along the beam path of:

$$z_f = \pm \, \pi \, \sqrt{(\rho^2 - 1)} \; (r_o^2/(2p + l + 1) \; \lambda)\qquad\qquad (2.15.)$$

where ρ = a factor determined by the application ≈1.05.

   r_o = beam radius (= 1.22F λ for a diffraction limited spot).

   p,l = Mode numbers.

Thus: $z = \pm \, \pi/\rho^2 - 1 \; (\, d_{min}^2/ \, 4 \, (\, 2p + l + 1) \; \lambda)$

$$z_f \cong 1.48 \; F^2 \lambda\qquad\qquad (2.16.)$$

Table 2.4 shows some figures for the focal spot size and the depth of focus given by different lenses with beams of different mode structures.

2.7. Optical Components

2.7.1. Lens Doublets

We have so far discussed the single simple lens. A doublet is an alternative to an aspheric lens for overcoming the effects of spherical

Table 2.4.	Effects of F number on the Focal Length and Depth of Focus for Different Mode Structures and Wavelengths.					
Wavelength	F Number (f/D)	Mode	Diffraction dmin mm	Depth of Focus zf mm	Spherical Aberration, dmin ** biconvex mm	planoconvex mm
10.6μm	2	top hat	0.26	0.08	0.5	0.36
	5	TEM00	0.13	0.5	0.08	0.06
	5	TEM01*	0.26	1.0		
	5	TEM20	0.65	2.5		
	10	TEM00	0.26	2.0	0.02	0.015
1.06μm	2	top hat	0.026	0.008	0.5	0.36
	5	TEM00§	0.013	0.05	0.08	0.06
	5	TEM01*§	0.026	0.1		
	5	TEM20§	0.065	0.25		
	5	Multi	~0.4	~2.0		
	10	TEM00	0.026	0.2	0.02	0.015

** For a beam diameter of 20mm and biconvex q=0; K = 0.1 and planoconvex q=1; K=0.073. (Where the value is smaller than the diffraction limited spot size, it means that spherical aberration is less significant. The two values should be added to obtain the approximate expected spot size.)

§ Current industrial high powered YAG lasers can not achieve this level of mode purity; but it can be seen that the incentive to do so is high.

Table 2.5	Comparison of Basic Lens Types(15)	
Type	Advantage	Disadvantage
Singlet	Low cost	High SA at low F No. no color correction
Air space aplanat (doublet)	Excellent TWD	Cost, no colour correction
Cemented achromatic doublet	Better SA than singlet	Low power only fair TWD for low F No.
Air space achromat (triplet)	Colour correction OK low F No. nearly diffraction limit at F/5	Cost
N.B. The colour correction is not relevant for single frequency lasers.		
SA = Spherical Aberration		
TWD = Transmitted Wavefront Distortion		

aberration. We have just noted that spherical aberration becomes the main issue for short focal length lenses of less than F/5. If such a short focus is needed then the doublet is the cheaper option to an aspheric lens. Table 2.5 shows a comparison of lens types.

The effect of a doublet compared to a singlet is illustrated in Fig. 2.19 (15). This figure illustrates the advantages to be found for doublets at low F numbers.

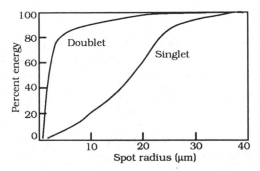

Fig. 2.19. Encircled energy at different radii for a singlet and doublet F/5 lenses focussing 1.06μm radiation from a Nd – YAG laser.

2.7.2. Depolarisers

When a polarised beam strikes a mirror surface at 45° to the plane of polarisation, the beam takes up two new planes: the p plane, parallel to the plane of incidence, and s plane, vertical to the plane of incidence. When this reflected beam strikes a mirror having a surface coating which is $\lambda/4$ thick in the direction of propagation of the beam, then the p-polarised beam (parallel to the plane of incidence) will penetrate the film while the s-polarised beam (perpendicular to the plane of incidence) will be reflected. The p-polarised beam will be reflected from beneath the film at the metal surface and so rejoin the main beam but it will be phase shifted by 2 [$\lambda/4$]. Thus the final beam will be one in which the plane of polarisation alternates between two states at right angles with every beat of the wave form. This gives the impression to a viewer from the end of the beam that the plane of polarisation is rotating. The beam is said to be "circularly" polarised. Some care has to be taken with these carefully designed coatings, which are usually of MgF_2, because they are slightly hygroscopic and cannot be safely wiped clean. Nevertheless depolarisers are now fitted to nearly all commercial cutting machines.

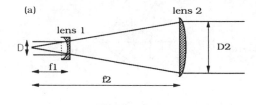

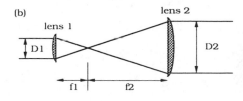

Fig.2.20. Examples of beam expanders and collimators. (a) Galilean; (b) Keplerian.

2.7.3. Collimators

A collimator or beam expander is often used in installations where the beam path is long or the laser produces such a small beam diameter that it is difficult to focus without having the lens very close to the work piece and therefore vulnerable to spatter. A transmissive beam expander is illustrated in Fig. 2.20 for the Galilean and Keplerian designs. The general principle is that the new beam size will be, $D_2 = D_1 f_2/f_1$. For long beam path work the beam divergence is one of the main criteria (16).

2.7.4. Metal Optics

2.7.4.1. Plane Mirrors:
The reflectivity of a mirror is a function of the material. Therefore most mirrors are made of a good conductor (good reflector) coated with gold for infrared radiation. The gold may be further coated with rhodium to allow gentle cleaning. New optics based on coated silicon are also used. In the case of the Laser Ecosse lasers these are sufficiently thin to allow gentle flexing. The reason for having good conductivity mirror substrates, apart from reflectivity, is the need for good cooling. This is usually achieved by water but may be by air blast.

The flatness of mirrors is achieved by careful machining. The most popular technique is single point diamond machining. The flatness is measured in an interferometer and recorded as $\lambda/*$, for example, $\lambda/5$ would mean that there was a variation in flatness of 1/5th of a wavelength over the mirror surface.

Cleaning mirrors is done by placing a soft lens tissue on the mirror and allowing a drop of methanol or isopropyl alcohol to fall on it then to draw the tissue over the face of the mirror until it is dry. This will prevent scratching and also drying stains. Never rub a mirror surface. If a mirror becomes tarnished or damaged in any way it is usually best to regrind and recoat it. If possible, mirrors should always be mounted so that they avoid dust falling on them.

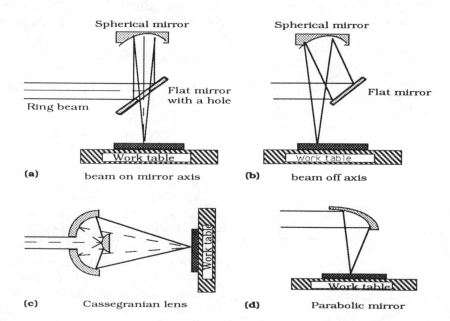

Fig. 2.21. Various ways of focussing using mirrors.

2.7.4.2. Metal Focussing Optics (Parabolic Mirrors): With the growing use of very high powered lasers with average powers over 5kW, transmissive optics are near the limit of their thermal stress resistance. Most operators of such equipment prefer to use metal optics for focussing, collimating and guiding. One focussing element which uses the least number of mirrors is an off axis parabolic mirror. The arrangement is illustrated in Fig. 2.21. They are very good if they are properly aligned, but they are very sensitive to alignment. Various arrangements are illustrated in Fig. 2.21.

2.7.4.3. Holographic Lenses: Reflecting plates finely etched or micro machined to two or three levels can be made in the form of a hologram and thus reflect an image of any required shape. The early versions, known as "kinoforms" (17) had a reflectivity of around 30% and there was some noise on the image at the edges. Modern versions are made of reflective material and are considerably more efficient.

2.7.4.4. Laser Scanning Systems (18): There are many occasions when a line beam is required. This can be achieved by a cylindrical lens or a scanning system. These scanning systems can be based on oscillating aluminium mirrors as shown in Fig. 2.22.a. These systems have the weakness of giving a nonuniform power distribution due to the turn point at the end of each oscillation. To avoid this a rotating polygon is

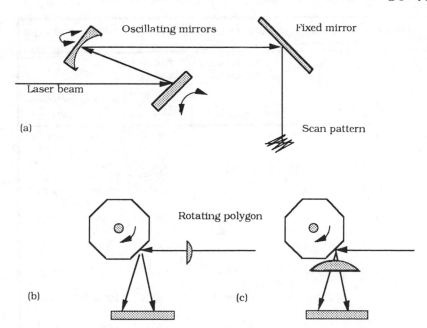

Fig. 2.22. Various scanning systems;
(a) Rastor system; (b) and (c) Rotating polygons

often used as shown in Fig. 2.22.b,c. This device has the problem of a varying velocity over the scan due to the varying angle. Zheng(19) developed a double polygon system which overcame that problem. There is some considerable geometry involved in designing these systems (18).

2.7.4.5. Fibre delivery systems (20): There are a variety of fibres being considered for delivering power beams for material processing. The advantages appear obvious by analogy with electricity. There are, however, some difficulties which need to be faced when delivering power down a fibre. The first is the problem with the insertion into the fibre. The fibres are often a fraction of a millimetre in diameter and thus the focussed beam is directed at the entry point into the fibre and any dirt will cause catastrophic absorption. Once in the fibre the intensities are, of course, very high - for example a 2kW beam in a 0.5mm diameter fibre would have an intensity of around 10^6 W/cm². Compare this value with the published values of damage thresholds for fibres shown in Table 2.6 (20) and the problem becomes apparent. If the fibre is made larger to reduce this value then the focussability is reduced and the major property of the laser beam is lost. The finest a beam from a multimode fibre can be focussed is as the image of the end of the fibre. This will never be anything like the fineness possible with a straight laser beam. The further problem is that of high loss due to nonlinear events such as Raman, Brillouin and Rayleigh scattering noted earlier.

Table 2.6.			Published values of the damage thresholds in various fibres (20)				
Pulse duration	Wavelength	Material	Transmission loss	Core diameter	Breakdown		
						power	intensity
	µm		dB/km	µm			W/cm2
40fs	0.620	SiO2	~7	3		150kW	>2 E12
5 ps	0.615	SiO2	20	3		1.5kW	2 E10
18ns	0.248	SiO2	2000	1000		16MW	>2 E9
100µs	1.06	SiO2	1	10		~500W	>5 E6
CW	1.06	SiO2	1	~10		100W	5 E6
500ns	2.94	ZrF4	12	100		~800W	>1 E7
CW	5.2	As2S3	900	700		100W	2.6 E4
CW	10.6	KRS-5	200-1000	250		20W	>4 E4
CW	10.6	hollow	1000	3000		800W	

There is thus a limit on the power transmission of high quality, high powered beams in fibres. One alternative could be a multiplicity of fibres as shown in Fig. 1.5 of three beams being focussed through one lens.

There is a rapidly growing market in fibre optic delivery systems for Nd-YAG lasers (21). The fibres are made from graded or stepped silica with the lower refractive index on the outside to keep the beam always refracting towards the centre, but plastic fibres are becoming of increasing importance. CO_2 laser radiation at 10.6μm has only been passed down special thallium based fibres with substantial losses reducing the permissible power down to around 100W. The fibre is also not particularly flexible. CO radiation at 5.4μm can be passed down CaF_2 or Zn halide fibres.

References

1.   Huygen.C. "Traite de la Lumiere" 1678, publ Leiden 1690.
2.   Newton.I. Opticks 1st edition 1704.
3.   Einstein.A. paper 1905; Nobel Prize 1921.
4.   Heisenberg.W.K. Nobel Prize 1932;   Schrodinger.E. Nobel Prize 1933.
5.   Cobine.J.D. "Gaseous Conductors" McGraw Hill, New York,1941.
6.   Nonhof.C.J. "Material Processing with Nd-YAG Lasers" Publ Elctro Chemical Publications Ltd., 8 Barns St., Ayr, Scotland, 1988.
7.   Hector.L.G., Kim.W.S., Ozisiki. "Propagation and reflection of thermal waves in finite mediums due to axisymmetric surface waves" Proc XXII ICHMT Int Symp. on Manufcaturing and Material Procesing, Dubrovnik, Aug 1990 to be publ.
8.   American Institute of Physics Handbook ed E.G.Gray McGraw Hill Book Co. 3rd Ed (1857).
9.   Jupner.W., Rohte.W., Sepold.G., Teske.K. DVS Berichte 63 p222 1983
10. Patel.R.S., Brewster.M.Q. "Effects of oxidation on low power Nd-YAG Laser metal interactions" Proc 7th Int Conf on Lasers and Electro Optics ICALEO '88, Oct/Nov 1988 Santa Clara Calif. publ Springer Verlag/IFS 1988 p313-323.
11. O'Neill.W. Ph.D. Thesis London University 1990.
12. Kielman.F. "Stimulated absorption of CO2 laser light on metals" Proc NATO Adv. Study Inst. on Laser Surface Treatment, San Miniato, Italy, Sept 1985 p17-22.
13. Jenkins.F.A., White.H.E. "Fundamentals of Optics" McGraw Hill Publishing Co London 2nd ed 1953.
14. Sharp.M., Henry.P., Steen.W.M., Lim.G.C. "An analysis of the effects of mode structure on laser material processing" Proc. Laser '83 Optoelectronic Conf. Munich June 1983 ed Waidelich. p243-246.

15. Lowrey.W.H., Swantner.W.H., "Pick a laser lens that does what you want it to" Laser Focus World May 1989 p121-130.
16. Zoske.U., Giesen.A. "Optimisation of beam parameters of focussing optics" Proc 5th Int Conf on Lasers in Manufacturing (LIM5) Stuttgart Sept.1988 publ IFS publications Ltd. p267-278.
17. Patt.P.J. "Binary phase gratings for material processing" Journ Laser Appl. Vol 2, No 2, p 11-17 1990.
18. Stutz.G.E. "Laser scanning systems" Photonics Spectra June 1990 p113-116.
19. Zheng.H.U. Ph.D. Thesis, London University 1990.
20. Weber.H.P., Hodel.W. "High power transmission through optical fibres for material processing" Ind. Laser Annual Handbook 1987 p33-39.
21. Walker.R. "Fibreoptic beam delivery leads to versatile systems" Industrial Laser Review ILR July 1990 p5-6.

"Now you know the difference between a moon beam and a laser beam!"

Chapter 3

Laser Cutting

"Measure a thousand times and cut once" Old Turkish proverb.

3.1. Introduction

The idea of cutting with light has appealed to many from the first time they burnt paper on a sunny day with the help of a magnifying glass. Cutting centimetre thick steel (Fig. 3.1) with a laser beam is even more fascinating!

Laser cutting is today the most common industrial application of the laser; in Japan around 80% of industrial lasers are used this way. Apart from the fascination, which is rarely a driving force for investment in hard industry, the reason is most probably that in cutting there is a direct process substitution into an established market and the laser, in many cases, happens to be able to cut faster and with a higher quality than the competing processes. The comparison with alternative techniques is listed in Table 3.1. The significant advantages of the laser are seen in the table but possibly these need some further explanation. Thus the advantages are:

Fig. 3.1. Laser cutting mild steel.

3.1.1. Cut Quality Characteristics

1. The cut can have a very narrow kerf width giving a substantial saving in material. (*Kerf* is the width of the cut opening.)
2. The cut edges can be square and not rounded as with most hot jet processes or other thermal cutting techniques.
3. The cut edge can be smooth and clean. The cut is reckoned to be a finished cut, requiring no further cleaning or treatment.
4. The cut edge is sufficiently clean it can be directly rewelded.
5. There is no edge burr as with mechanical cutting techniques. Dross adhesion can usually be avoided.
6. There is a very narrow HAZ (Heat Affected Zone), particularly on dross free cuts. Usually there is a very thin resolidified layer of micron dimensions. Thus there is negligible distortion.
7. Blind cuts can be made in some materials, particularly those which volatilise, such as wood or acrylic.
8. Cut depth is limited and depends on the laser power. 10-20mm is the current range for high quality cuts.

3.1.2. Process Characteristics

9.   It is one of the faster cutting processes.
10. The work piece does not need clamping, though this is usually advisable to avoid shifting with the table acceleration and for locating when using a CNC program.
11. Tool wear is zero since the process is a non contact cutting process.
12. Cuts can be made in any direction; but note section on polarisation (3.5.1.3).
13. The noise level is low.
14. The process can be easily automated with good prospects for adaptive control in the future.
15. Tool changes are mainly "soft". That is they are only programming changes. Thus the process is highly flexible.
16. Some materials can be stack cut, but there may be a problem with welding between layers.
17. Nearly all materials can be cut. They can be friable, brittle, electric conductors or non conductors, hard or soft. Only highly reflective materials such as aluminium and copper can pose a problem but with proper beam control these can be cut satisfactorily.

3.2. The Process - How It Is Done

The general arrangement for cutting with a laser is shown in Fig. 3.2.

Table 3.1. Comparison of Different Cutting Processes

QUALITY	Laser	Punch	Plasma	Nibbling	Abrasive Fluid Jet	Wire EDM	NC Milling	Sawing	Ultrasonic	Oxy Flame
Rate	✓	✓	✓	✗	✗	✗	✗		✗	✗
Edge Quality	✓	✓	✗	✗	✓	✓	✓	✗	✓	✗
Kerf Width	✓	✓	✗		✓	✓		✗		
Scrap and Swarf	✓	✓		✗	✓			✗		✓
Distortion	✓		✗		✓		✓		✓	✗
Noise	✓	✗	✗		✗					✗
Metal+Nonmetal	✓		✗		✓		✓		✓	
Complex Shapes	✓	✗	✓							
Part Nesting	✓	✗			✓					
Multiple Layers	✗	✓								
Equipment Cost	✗				✗			✓		✓
Operating Cost						✗		✓		✓
High Volume	✓	✓		✗					✗	
Flexibility	✓	✗	✓	✓	✓	✗				
Tool Wear	✓	✗	✓	✗	✓		✗	✗	✗	✓
Automation	✓	✓	✓	✗	✓	✗	✓			✗
HAZ	✓	✓	✗		✓	✓	✓			✗
Clamping	✓	✗	✓		✓		✗	✗		
Blind Cuts	✓	✓	✓	✓		✗	✗		✗	✓
Weldable Edge	✓	✓	✗	✓	✓	✓	✓		✓	✗
Tool Changes	✓	✗	✓		✓					

✓ Point of particular merit
✗ Point of particular disadvantage
(Further comparisons can be found in ref 1)

The principle components are the laser itself with some shutter control, beam guidance train, focussing optics and a means of moving the beam or workpiece relative to each other. The shutter is usually a retractable mirror which blocks the beam path and diverts the beam into a beam dump which doubles as a calorimeter. When the beam is required the mirror is rapidly removed by a solenoid or pneumatic piston. The beam then passes to the beam guidance train which directs the beam to centre on a focussing optic. The focussing optic can be either transmissive or reflective; the transmissive optics are made of ZnSe, GaAs or CdTe lenses for CO_2 lasers or quartz lenses for YAG or excimer lasers; the reflective optics consist of parabolic off axis mirrors (see Section 2.7). The focussed beam then passes through a nozzle from which a coaxial jet flows. The gas jet is needed both to aid the cutting operation and also to protect the optics from spatter. In the case of the metal optics system,

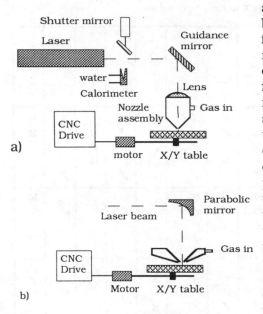

a)

b)

Fig. 3.2. General arrangement for laser cutting. a) transmissive optics; b) reflective optics.

an "air knife" is often used which blows sideways across the exit from the optic train thus deflecting any smoke and spatter. For cutting processes which rely on melt removal by the gas jet there is a problem for the metal optics systems. The main reason for using metal optics is either that one runs a research school with clumsy students or that the power of the laser is such that the transmissive optic is near the limit of its thermal stress tolerance. In fact if the transmissive optics are likely to break then one uses metal optics. To achieve a gas jet suitable for cutting (>20m/s and reasonably well focussed) without interposing a transmissive element, a set of centrally directed nozzles (2) or a ring jet can be used. It is very important to the process that the beam, optic and jet are all lined up.

3.3. Methods of Cutting

This general arrangement can be used to cut in six different ways, shown in Table 3.2.

3.3.1. Vaporisation Cutting

In cutting which relies on vaporisation, the focussed beam first heats up the surface to boiling point and so generates a keyhole. The keyhole causes a sudden increase in the absorptivity due to multiple reflections and the hole deepens quickly. As it deepens so vapour is generated and escapes blowing ejecta out of the hole or kerf and stabilising the molten walls of the hole (3). This is the usual method of cutting for pulsed lasers or in the cutting of materials which do not melt such as wood, carbon, and some plastics.

The rate of penetration of the beam into the workpiece can be estimated from a lumped heat capacity calculation assuming the heat flow is one dimensional and all of it is used in the vaporisation process - that is that

Table 3.2.	Different ways in which the laser can be used to cut.	
Method	Concept	Relative Energy
1. Vaporisation		40
2. Melt and blow		20
3. Melt, burn and blow		10
4. Thermal stress cracking	Tension	1
5. Scribing		1
6. "Cold cutting"	$h\nu$ High energy photons	100

Table 3.3.	Material properties and penetration speeds, V, and time to vaporise, Tv, for a beam of power density 6.3 x E10 W/m2 (4,5).								
	Material Properties							Process properties	
Material	ρ	Lf	LV	Cp	Tm	Tv	K	V	tv
	kg/m3	kJ/kg	kJ/kg	J/kgC	C	C	W/mK	m/s	μs
Tungsten	19300	185	4020	140	3410	5930	164	0.64	3
Aluminium	2700	397	9492	900	660	2450	226	1.9	0.6
Iron	7870	275	6362	460	1536	3000	50	1.0	0.3
Titanium	4510	437	9000	519	1668	3260	19	1.2	0.09
Stainless steel (304)	8030	~300	6500	500	1450	3000	20	0.97	0.4

the heat conduction is zero. This fairly gross assumption is not ridiculous if the penetration rate is similar to or faster than the rate of conduction.

Thus the volume removed per second per unit area = penetration velocity, V m/s.

$$V = F_o/\rho[L + C_p(T_v-T_o)] \qquad (m/s) \qquad (3.1)$$

where:

F_o	= Absorbed power density	(W/m^2)
ρ	= Density of solid	(kg/m^3)
L	= Latent heat of fusion and vaporisation	(J/kg)
C_p	= Heat capacity of solid	$(J/kg°C)$
T_v	= Vaporisation temperature	$(°C)$
T_o	= Temperature of material at start	$(°C)$

If we substitute values into this equation we can derive the approximate maximum penetration rate possible for different materials.

Assuming we have a 2kW laser focussed to 0.2mm beam diameter the power density will be:

$$F_o = 2000 \times 10^6/\pi 0.1^2$$
$$= 6.3 \times 10^{10} \qquad (W/m^2)$$

The penetration rate of such a beam into various materials is calculated in Table 3.3. These penetration figures are of the same order as those found experimentally (4). If the penetration rate is around 1m/s then the vapour velocity from a cylindrical hole would be ρ_v/ρ_s = 1000m/s. At these sonic speeds compression effects and variations in the hole shape will mean the actual velocity of exit of the vapour is much less, but nevertheless sonic flow and shock waves will occur and the flow will be capable of considerable drag in eroding the walls of the hole formed. Thus in this form of cutting the material is removed part as vapour and part as ejecta. Gagaliano and Paek (1971) (6) estimated from their experiments that around 60% of the material was removed as ejecta.

The quality of the hole or cut is determined by the quantity of melt which may build up and cause debris on the surface or erosion marks on the wall. Thus it is interesting to calculate how quickly the boiling point is reached and so see how the melt may be reduced.

For one dimensional heat flow with constant energy input it can be shown (see Chapter 5) that the surface temperature at any time, t, after the start of irradiation is given by:

$$T(0,t) = (2F_o/K)[(\alpha t)/\pi]^{1/2} \qquad\qquad (3.2.)$$

where:
α = thermal diffusivity $(K/\rho C_p)$ $\qquad$ (m²/s)
K = thermal conductivity $\qquad\qquad$ (W/mK)

Therefore: $t_v = \pi/\alpha \, [(T_B K)/(2F_o)]^2$

The estimated time for a 2kW laser beam to cause vaporisation is shown in Table 3.3. The thermal gradient at that time would have penetrated (assuming Fourier number $(x^2/\alpha t) = 1$) around 2μm for iron and hence it can be seen that the HAZ is expected to be small in this case.   The previous calculation, based upon ignoring the heat conduction, is thus not so wide of the mark.

From this calculation it can be seen that the peak power of the beam is very important. In drilling with a YAG laser considerable attention is paid to the designing of the pulse shape with time. The aim is usually to have a short sharp pulse for cutting as opposed to a longer pulse with reduced initial peak for welding. The shaping of the pulse can be critical, as in the drilling of glass (7).

There are a number of side effects from this almost explosive evaporation. One is the recoil pressure required to accelerate the vapour away. Bernoulli's equation is able to give a rough estimate of the value of this pressure for an exit velocity of 1000m/s even though it assumes incompressible flow, as:

$$\Delta P = \rho_v v^2/2 = 4 \times 10^{6}\,N/m^2$$

One atmosphere is $10^5\,N/m^2$. A pressure rise of this order will cause a rise in the vaporisation temperature. This pressure causes stress in the surface. This stress is amplified by the thermal stresses generated in the heated surface. Together they represent quite a considerable stress.   If this can be applied very quickly, in a few nanoseconds (s^{-9}) then the effect is similar to being hit, as in shot peening. Such a process is known as laser shock hardening and is mentioned in Chapter 6.

3.3.2. Fusion Cutting - Melt and Blow

Once a penetration hole is made or the cut is started from the edge, then it is possible with a sufficiently strong gas jet to blow the molten material out of the cut kerf and so avoid having to raise the temperature to the boiling point or any further. It is thus not surprising to find that cutting in this manner requires only one tenth of the power for vaporisation cutting. Note the ratio of the latent heat of fusion for melting and boiling in Table 3.3.

The process can be approximately modelled by assuming all the energy enters the melt and is removed before significant conduction occurs. Once more, since the HAZ for good cuts by this method rarely exceeds a few microns this assumption is not so daft.

We thus have a simple lumped heat capacity equation based on the heat balance on the material removed similar to equation (3.1.) as shown in the Fig. 3.3.

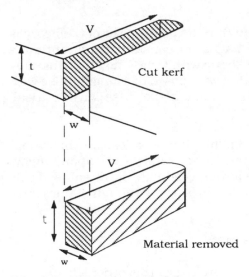

The balance is:
$$\eta P = wtV.\rho[C_p\Delta T + L_f + m'L_v]$$

where:
P   = Incident power   W
w   = Average kerf width m
t   = Thickness   m
V   = Cutting speed   m/s
m' = Fraction of melt vaporised
L_f = Latent heat of fusion J/kg
L_v = Latent heat of vaporisation
                                    J/kgK
ΔT =Temperature rise to cause
          melting   K
η = Coupling coefficient
ρ = density   kg/m³

Fig. 3.3. Volume melted and removed during cutting.

Rearranging this equation we get:

$$[P/tV] = w\rho/\eta\{\ C_p\Delta T + L_f + m'L_v\} = f(material)\quad J/m^2 \qquad (3.3.)$$

Apart from the value of the kerf width, w - a function of spot diameter and to some extent speed - and the coupling efficiency, η, the other variables are all material constants. Thus it is reasonable to expect that the group [P/tV] is constant for the cutting of a given material with a given beam. A collation of the data from the literature is presented in Figs. 3.4,3.5,3.6. The straight line correlation is significant considering all the unspecified different cutting methods used by the various authors - particularly in regard to the effects of polarisation and gas jets

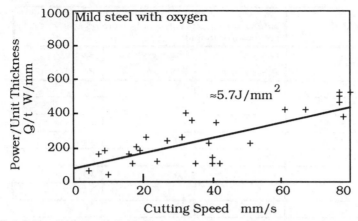

Fig. 3.4. P/t vs V for mild steel.

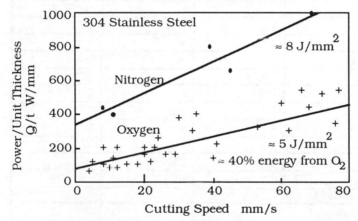

Fig. 3.5. P/t vs V for stainless steel.

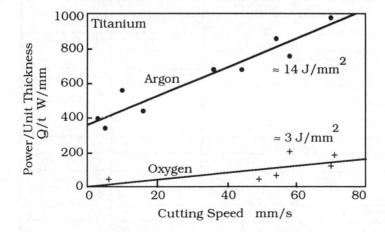

Fig. 3.6. P/t vs V for titanium.

Table 3.4	Average severance energies for CW CO2 laser cutting found experimentally from a variety of sources (principally 8,9).		
Material	Lower Value P/Vt J/mm2	Higher Value P/Vt J/mm2	Average P/Vt J/mm2
Mild Steel + O2	4	13	5.7
Mild Steel + N2	7	22	10
Stainless Steel + O2	3	10	5
Stainless Steel + Ar	8	20	13
Titanium + O2	1	5	3
Titanium + Ar	11	18	14
Aluminium + O2			14
Copper + O2			30
Brass + O2			22
Zirconium + O2			1.7
Acrylic Sheet	1	3	1.2
Polythene	2.7	8	5
Polypropylene	1.7	6.2	3
Polystyrene	1.6	3.5	2.5
Nylon	1.5	5	2.5
ABS	1.4	4	2.3
Polycarbonate	1.4	4	2.3
PVC	1	2.5	2
Formica	51	85	71
Phenolic Resin			2.7
Fibre Glass(epoxy)			3.2
Wood: Pine(yellow)			23
Oak			26
Mahogany			24
Chipboard	45	76	59
Fibreboard			50
Hardboard			23
Plywood	20	65	31
Glass			20
Alumina	15	25	20
Silica			120
Ceramic Tile			19
Leather			2.5
Cardboard	0.2	1.7	0.5
Carpet (auto)			0.5
Asbestos Cement			5.0

N.B. These figures do not apply to Nd-YAG pulse cutting where the mechanism is different: for example for mild steel Nd-YAG values are between 15-200J/mm2

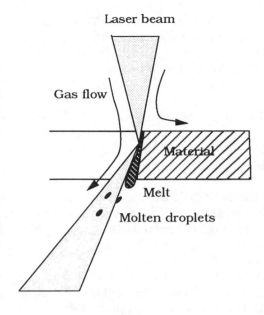

Laser beam

Gas flow

Material

Melt

Molten droplets

Fig. 3.7. Interactions at the cutting front.

as discussed later. So it is possible to draw up a chart of the severance energy (J/mm²) required for unit area severed for different materials, as in Table 3.4.

This at least establishes the main operating parameters. What is actually happening at the cutting front is of considerable complexity. Fig. 3.7 shows the cut front in section. The beam arrives at the surface and most passes into the hole or kerf; while some is reflected off the unmelted surface, some may pass straight through. At slow speeds the melt starts at the leading edge of the beam and much of the beam passes clean through the kerf without touching the material if it is sufficiently thin (10). The absorption takes place on the steeply sloped cut front (θ = approx 14° to the vertical (11)) by two mechanisms: mainly by Fresnel absorption - that is direct interaction of the beam with the material - and secondly by plasma absorption and reradiation. The plasma build up in cutting is not very significant due to the gas blowing it away. Thus the power density on the front is $F_o \sin\theta = F_o$ x 0.24. This causes melting and the melt is then blown away by the drag forces from the fast flowing gas stream. At the bottom of the kerf the melt is thicker due to deceleration of the film and surface tension retarding the melt from leaving. The gas stream ejects the molten droplets at the base of the cut into the atmosphere. In driving through the kerf, the gas would entrain the surrounding gas in the kerf and generate a low pressure region further up the cut length. This can have a detrimental effect by sucking the dross back into the cut. In fact the problem of the removal of the dross from the bottom edge is further complicated by the wetability of the workpiece to the melt and the flow direction of the gas jet. Thus cutting thin tin plate is difficult due to the dross clinging to the molten tin plate and the poorly directed gas jet which is emitted from a slot in thin material. The gas stream not only drags the melt away but will also cool it. Infact both momentum and heat transfer will occur. The extent of the cooling can be calculated. The heat removal by convection is described by: $Q = hA(\Delta T)$

an equation which defines the heat transfer coefficient, h. The value of h has been determined for many geometries and is given in (12). It is usually quoted as Nu = f(Re,Pr). An approximate, and high, estimate of h can be derived, such as h < 100W/m²K. The heat loss at the cut front now becomes: $Q = 100 \times t \times w \times \Delta T$

For a thickness, t, of 2mm and kerf width, w, of 1mm and the ΔT of 3000K - all high values - then Q = 0.6W. Thus the cooling effect of the gas is negligible compared to the few 1000W being delivered by the beam, mainly due to the small area involved in laser cutting.

In fusion cutting the action of the gas is to drag the melt away and little else. The design of the nozzle and the alignment of the nozzle with the laser generated kerf are important areas of concern in as much as they affect the drag of the gas on the melt.

As the cut rate is increased the beam is automatically coupled to the work piece more efficiently by less being lost through the kerf (10). Also the beam tends to ride ahead onto the unmelted material. When this occurs the power density increases since the surface is not sloped and so the melt proceeds faster and is swept down into the kerf as a step. As the step is swept down it leaves behind a mark on the cut edge called a striation (see Fig. 3.8). The cause of striations is a subject of some dispute, there are many theories: the step theory just outlined, the critical droplet size causing the melt to pulsate in size before it can be blown free (13) and the sideways burning theory. There are conditions under which no striations occur. These are governed by gas flow or by pulsing at the frequency of the natural striation (14). A further feature of the cut face is that there is often, but not always a break in the flow lines . This may be due to the start of the first reflection of the beam off the cut face, the end of the burning reaction (see next section) or it may be a shock wave phenomenon. Currently this is not well understood.

3.3.3. Reactive Fusion Cutting

If the gas in the previous method is also capable of reacting exothermically with the workpiece then another heat source is added to the process. Thus the cut front becomes an area of many activities. Fig. 3.7 illustrates the basic structure of the cut face. The gas passing through the kerf is not only dragging the melt away, as just seen, but is also reacting with the melt. Usually the reactive gas is oxygen or some mixture containing oxygen. The burning reaction starts usually at the top when the temperature reaches the ignition temperature. The oxide is formed and is blown into the kerf and will cover the melt lower down.

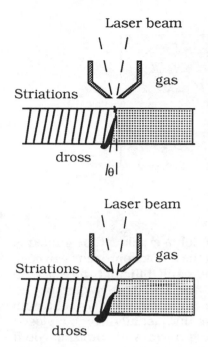

Fig. 3.8. The stepwise formation of striations.

This blanketing will slow the reaction, and may even be the cause of the break in the striation lines just noted. It can be seen from Figs. 3.5 and 3.6 that the amount of energy supplied by the burning reaction varies with the material; with mild steel it is 60%; with stainless steel it is also 60% and with a reactive metal like titianium it is around 90%. Thus cutting speeds are usually at least doubled using this technique. As a general rule: the faster the cut, the less heat penetration and the better the quality. However since there is a cutting reaction taking place some chemical change in the workpiece may be expected. With titanium this can be critical since the edge will have some oxygen in it and will be harder and more liable to cracking. With mild steel there is no noticeable effect except a very thin resolidified layer of oxide on the surface of the cut. An advantage is that the dross is no longer a metal but is usually an oxide which for mild steel flows well and does not adhere to the base metal as strongly as if it were metal. With stainless steel the oxide is made up of high melting point components such as Cr_2O_3 (melting point~2180°C) and hence this freezes quicker causing a dross problem. It is the same with aluminium.

Due to the burning reaction a further cause of striations is introduced. In slow cutting, at speeds less than the burning reaction, the ignition temperature will be reached and then burning will occur proceeding outward in all directions from the ignition point as illustrated in Fig. 3.9. This mechanism is only plausible as a cause for striations if the cut is slow. In this case very coarse striations are revealed as illustrated in the slow cut of Fig. 3.10.

3.3.4. Controlled Fracture

Brittle material which is vulnerable to thermal fracture can be quickly and neatly severed by guiding a crack with a fine spot heated by a laser. The laser heats a small volume of the surface causing it to expand and hence to cause tensile stresses all around it. If there is a crack in this space, it will act as a stress raiser and the cracking will continue in the

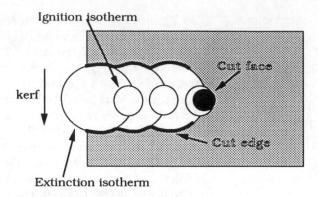

Fig. 3.9. Striation formation due to sideways burning.

direction of the hot spot. The speed at which a crack can be guided is swift of the order of m/s. This is fine until the crack approaches an edge when the stress fields become more complex and difficult to forecast. As a cutting method for glass it is superb. The speed, edge quality and precision are very good. The only problem is that for straight cuts snapping is quicker and for profiled cuts one usually needs a closed shape, as for the manufacture of car wing mirrors. If someone could solve the control problem on completing the closed form then a significant process would have been developed.

This process requires that the surface is not melted or it may damage the edge. It thus requires very little power. Typical figures are shown in Table 3.5.

3.3.5. Scribing

This is a process for making a groove or line of holes either fully penetrating, or not, but sufficient to weaken the structure so that it can be mechanically broken. The quality, particularly for silicon chips and alumina substrates, is measured by the lack of debris and low heat affected zone. Thus low energy, high power density pulses are used to remove the material principally as vapour.

3.3.6. Cold Cutting

This is a new technique only recently observed with the introduction of high powered Excimer lasers working in the ultraviolet. The energy of the ultraviolet photon is reported as 4.9eV in Table 2.1. This is similar to the bond energy for many organic materials. Thus if a bond is struck by such a photon then it may break. On the whole if it did, it would recombine and no one would be any the wiser. However once the dice

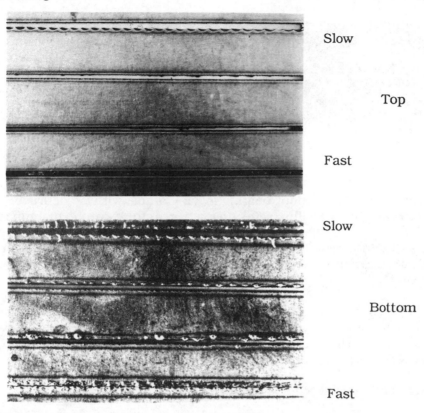

Slow

Top

Fast

Slow

Bottom

Fast

Fig. 3.10. Top and underside views of cuts in mild steel made at various speeds. The coarse striations formed at low speeds are clearly seen.

Table 3.5	Controlled Fracture Cutting Rates			
Material	Thickness mm	Spot Diameter mm	Incident Power W	Rate of separation m/s
99% Al2O3	0.7	0.38	7	0.3
	1.0	0.38	16	0.08
Soda Glass	1.0	0.5x12.7	10	0.3
Sapphire	1.2	0.38	12	0.08
Quartz (cryst)	0.8	0.38	3	0.61

Fig.3.11. A human hair carved using an excimer laser.

is rolled it could recombine in another way as with sun bathing and the generation of tan (or carcinogens!). Ultraviolet light is just at the beginning of the biologically hostile radiation range which goes on into X rays and gamma rays. When this radiation is shone onto plastic with a sufficient flux of photons that there is at least one per bond (15) then the material just disappears without heating leaving a hole with no debris or edge damage. Fig. 3.11 illustrates how a human hair can be machined.

This exciting new technique seems to be a dream come true for the electronic manufacturer and certainly that industry is not slow to take it up (16). Though marking is also an attractive application.

The potential medical applications include a dazzling array of possibilities in micro surgery and engineering with single cells, as well as more conventional tumour ablation. The scale and power range are, however, outside the interests of this book.

3.4. Theoretical Models of Cutting

The simple model presented in Section 3.3 covers a surprising amount of detail in describing the laser cutting process. For a more detailed analysis care is taken over describing the heat flow into the cut face as a line source (13) or as a cylindrical source (17). Analytic models, however, are limited in their ability to model detail in real world problems. Thus numerical models have been attempted and some useful semiquantitative models have been developed. These models are discussed in Chapter 5.

3.5. Practical Performance

Laser cutting is a multiparameter problem and hence sometimes difficult to understand regarding the interrelationship between all the parameters. The parameters can be grouped as:

Beam properties: spot size and mode,
                          power, pulsed or CW
                          polarisation
                          wavelength

Transport properties: speed
focal position

Gas properties: jet velocity
nozzle position, shape, alignment,
gas composition

Material properties: optical
thermal

3.5.1. Beam Properties

3.5.1.1. Effect of Spot Size: The principle parameters are laser power, traverse speed, spot size and material thickness as seen in the simple model, Section 3.3. Quite the most important of these is the spot size. This acts in two ways; firstly, a decrease in spot size will increase the power density which affects the absorption and secondly, it will decrease the cut width. Lasers with stable power and low order modes - usually true TEM_{00} modes, as opposed to irregular mountain modes! - cut considerably better than other lasers. Fig. 3.12. (2) shows the effect of mode on cutting performance and the results of Sharp (18) Fig. 3.13 in cutting mirrors is not possible with any other form of beam. Notice with Sharp's work that he was using only relatively little power to cut 0.5cm thick gold plated copper mirrors! In the hole drilling work of Shaw (19), with holes of aspect ratio of 100 using only 100W of power, he attributes this amazing performance to a very low order mode YAG laser with low power to avoid explosive erosion effects. Poor mode structures tend to produce cuts which compare with a good plasma torch. The spot size is controlled by the laser design - which establishes the mode and the optics which decides how fine the focus will be - see Section 2.6.1. Usually a lens F no. of around 5 is selected.

3.5.1.2. Effect of Power: The overall effect of the power is to allow cutting at faster speeds as shown in Figs. 3.4, 3.5 and 3.6. Some further striking results in cutting have been obtained by pulsing particularly if the pulse is extra high as in the latest super or hyper pulse lasers. These power spikes give enhanced penetration. The speed is controlled by the need to overlap the pulses.

3.5.1.3. Effect of Beam Polarisation: Fig. 3.14 illustrates the problem. The maximum cutting speed is doubled, cutting in one direction as opposed to one at right angles when cutting with a plane polarised laser beam. Nearly all high powered lasers have folded cavities which favours the amplification of radiation whose electric vector is at right angles to the plane of incidence to the fold mirrors. That is a horizontal folding will

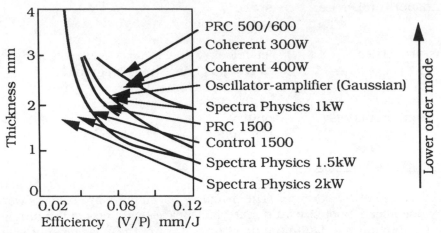

Fig. 3.12. The effect of mode on the cutting performance (2).

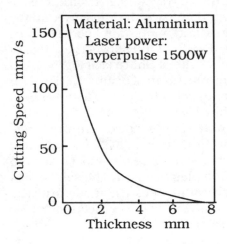

Fig. 3.13. Cutting mirrors (18).

produce a beam polarised vertically. If the cavity is not folded or the folding has near normal reflections then the beam will still be plane polarised but the plane of polarisation may move unpredictably with time. This is serious in view of Fig. 3.14. So even these lasers are now equipped with a fold at the total reflecting mirror in order to stabilise the plane of polarisation from the cavity. The cause of the phenomenon shown in Fig. 3.14 is that at the cutting face there is a glancing angle of incidence and as observed in Chapter 2 there is a distinct difference in the reflection of a beam at these angles depending upon whether the electric vector is at right angles to the plane of incidence (s-polarisation) or in the plane of polarisation (p-polarisation). If it is s-polarised then it will suffer a high reflectivity as shown in Fig. 2.9. If p-polarised it will be preferentially absorbed. This can be imagined as due to the form of oscillation expected of the interacting electron, as discussed in Section 2.5.4. This polarisation phenomenon was first noted by Olsen (20). Since then nearly all production cutting machines have been fitted with a circular polariser described in Chapter 2 Section 2.7.2. Such beams cut equally well in all directions and with a performance between that of the two plane polarised beams.

3.5.1.4. Effect of Wavelength: The shorter the wavelength the higher the

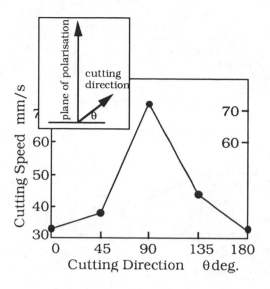

Fig.3.14. The effect of polarisation on the cutting performance with direction of cut.

absorptivity (see Chapter 2 Section 2.3.1), and the finer the focus for a given mode structure and optics train (see Section 2.6). Thus YAG radiation is preferable to CO_2 radiation as a general rule, though due to the poor mode structure of most YAG lasers of any significant power, the spot sizes for both CO_2 and YAG are similar, with an advantage for the true TEM_{00} laser. However, some curious results are found in the cutting of stainless steel and plastic with CO radiation at 5μm, shown in Fig. 3.15 (21). This is not understood at present and the results are under some discussion.

However the results from the Culham Laboratories on plastics (22) were done with the same resonator and optics and so represent a fairly good comparison between CO_2, 10.6μm radiation and that at 5.4μm from a CO laser. It is felt that the difference lies in the absorption of the beam on the cut face. As

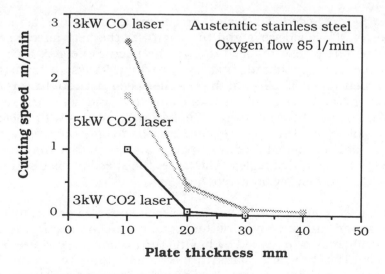

Fig. 3.15. Comparison of CO and CO laser cutting (21).

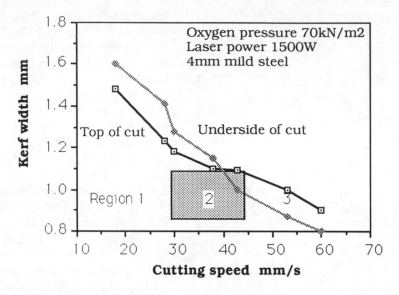

Fig. 3.16. Variation of Kerf Width with Cutting Speed (23).

oxidation occurs so a thin film is formed which may cause enhanced absorption for wavelengths for which the film is equal to a quarter wavelength, $\lambda/4$. The effect of wavelength is hard to imagine other than in the absorption and focussability.

3.5.2. Transport Properties

3.5.2.1. Effect of Speed: The faster the cutting the less time there is for the heat to diffuse sideways and the narrower the heat affected zone (HAZ). The kerf is also reduced due to the need to deposit a certain amount of energy to cause melting. Hence with a Gaussian beam there is a "sharpened pencil" effect in that as the speed rises so there is only sufficient energy at the tip of the Gaussian curve and not at the root to cause melting and hence cutting. The kerf width varies with speed as shown in Fig. 3.16. The three regions are due to side burning at slow speeds, stable cutting at medium speeds and failure for the dross to clear in the higher speed region. The faster the speed the better the cut finish until this last region is reached.

3.5.2.2. Effect of Focal Position: The surface spot size determines the surface power intensity and whether penetration will occur but optimum cutting may be obtained by having the minimum spot size below the surface. The problem is related to absorption on the cut face and how to keep the energy together (see discussion in Section 4.4.6). Very deep

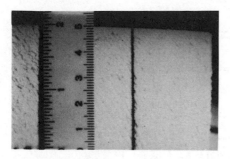

Fig. 3.17. A laser cut through 5cm of block board.

cuts are rarely achieved with any great quality since the beam spreads out and suffers multiple reflections. There are exceptions; consider Fig. 3.17, a 5cm cut with parallel walls in block board. How could the beam do that? It must have been wave guided down a slot whose walls are made of graphite - not a normal material to consider for reflections!

3.5.3. Gas Properties

3.5.3.1. Effect of Gas Jet Velocity: It has been described how the gas jet operates by dragging the melt out of the cut. The quicker it can be removed the quicker the next piece can be melted. Thus since the drag depends on the slot Reynolds number ($\rho u d/\mu$) the velocity of the gas in the slot on the cut face is critical. Gabzdyl (2) made some experiments by directing the jet at various angles into the cut front with little effect. Increasing the gas jet velocity increased the cutting rate up to a point as seen in Fig. 3.18. It was a puzzle as to why there should be this fall off in cutting speed with nozzle pressure. Some early workers suggested cooling was the problem but the calculation in Section 3.3. showed this to be incorrect or at least very unlikely. Kamalu (24) performed some schlieren experiments and showed that there was a density gradient field (DGF) adjacent to the cut surface which could be affecting the focus at the cut front. However the DGF was like a lens and hence the effect was difficult to justify to the extent shown in Fig. 3.18. Brook Ward (25) at the Culham Laboratories took some surface pressure measurements and showed that there was a series of shock phenomena associated with the high pressure jets. The structure is illustrated in Fig. 3.19. The result was plotted by him as pressure fields, Fig. 3.20. The first Mach shock disc is expected to influence cutting when the nozzle distance is around 2mm - the very distance at which most people were working! Using his pitot system he worked through a maze of nozzle shapes in the hope of finding the best shape to avoid this problem. It was shown that a nozzle having an orifice with an odd number of lobes, such as "☆", avoided the shock disc problem. However it introduced the further problem of how to keep such a shape when there is the chance of beam clipping or simple back reflection which might damage the nozzle. Undeterred by all this fine science the job shop users were happily building nozzles to go to even higher pressures and cutting with nozzle pressures of 14bar or so using specially designed optics to withstand the pressure, Fig. 3.21 (26). This has been found to have beneficial effects

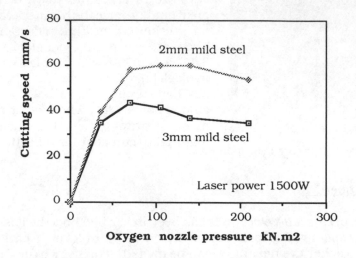

Fig. 3.18. Variation of cutting speed with pressure (24).

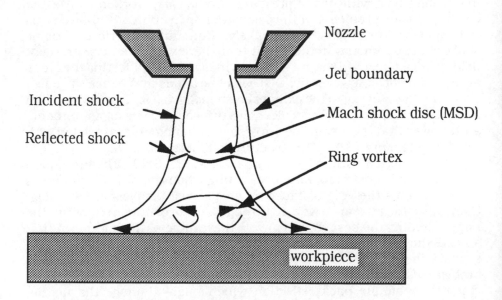

Fig. 3.19. The structure of an impinging sonic jet (25).

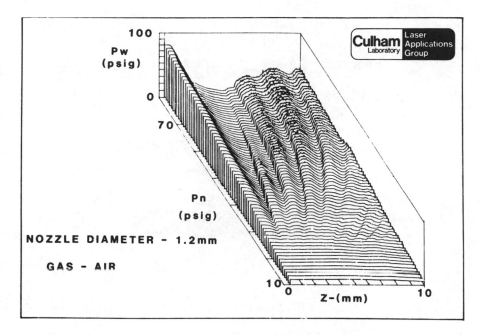

Fig. 3.20. Pressure field on plate from an impinging jet for various distances and pressures(25).

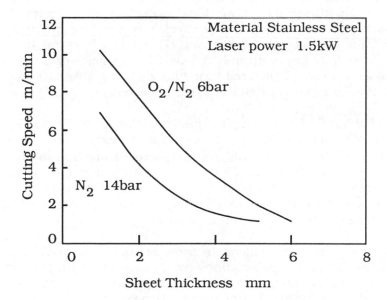

Fig. 3.21. Cutting stainless steel with inert gas and high pressures (26).

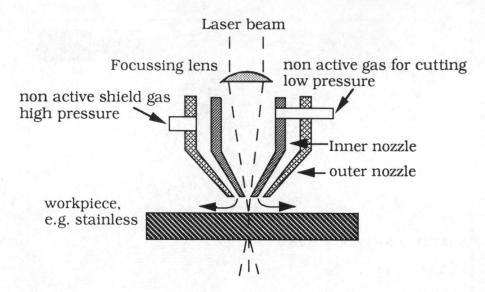

Fig. 3.22. A High Pressure Ring Nozzle used for "Clean Cut" Technique (27).

which are not in the public literature at the time of writing. Multiple nozzle systems are being used in several areas. The so called "clean cut" nozzle of Amada shown in Fig. 3.22 (27) operates at around 1 atmosphere pressure on the inner jet for lens protection and around 5 atmospheres in the outer ring jet. The effect is to produce burrless, striation free cuts. For example, Mitsubishi (27) claims to have cut 4mm Al (A5052 Al 2.4%Mg alloy) with only 1.8kW CO_2 laser power by this method. Amada have also pioneered a ring nozzle using a water spray on the outer nozzle which is intended to reduce dross and HAZ.

3.5.3.2. *Effect of Nozzle Alignment:* The quality of the cut is affected by the alignment of the nozzle with the laser beam. An exhaustive set of experiments was undertaken by Gadzdyl (28) methodically misaligning the beam and jet. This alignment affects both the roughness of the cut and the way the dross clears the kerf; for example it is possible to deliberately misalign the beam to make the specimen clear of dross but with all the dross clinging to the waste material.

3.5.3.3. *Effect of Gas Composition:* In some results from BOC Ltd., Zheng (29) showed that the gas composition has an effect on the cut quality. Remembering that the quality rises with speed those mixtures which allow faster cutting give better results. However reactive cutting has a greater tendency to produce striations. The effect is not very strong but certain mixtures are better than others. Pure oxygen is good but with a flavour of He it is sometimes better. This area is not one which has

Table 3.6.	Effects of Surface Treatment on Cutting Speeds (30)					
	Polished		Untreated		Shot blasted	
Material	Vel mm/s	Power W	Vel mm/s	Power W	Vel mm/s	Power W
C263 Ni Alloy	12.7	600	12.7	600	21.1	600
N80 Ni Alloy	12.7	400	16.9	400	21.1	400
L2%Cr Steel	12.7	200	25.4	200	25.4	200

Table 3.7	Behaviour of Different Materials to Laser Cutting
Property	Material
High Reflectivity (Need for Fine Focus)	Gold, Silver, Copper, Aluminium, Brass
Medium/High Reflectivity	Most metals
High Melting Point	W, Mo, Cr, Ta, Ti, Zr
Low Melting Point	Fe, Ni, .Sn, Pb
High Oxide Melting Point (Dross Problems)	Cr, Al, Zr
Low Reflectivity	Most non metals
Organics	
Tendency to char	PVC, Epoxy, Leather, Wood, Rubber, Wool, Cotton
Less tendency to char	Acrylics, Polythene, Polypropylene, Polycarbonate
Inorganics	
Tendency to crack	Glass, Natural Stones
Less tendency to crack	Quartz, Alumina, China, Asbestos, Mica
See also list of the cuttability of many materials in Industrial Laser Annual Handbook 1990 pp3-6, published Penwell Books, Tulsa, Oklahoma,USA.	

been explored in detail to date, but it may be that there is an optimum burning rate for a given cutting speed to give good edge quality. Thus the gas mixture might advantageously be arranged to give the matching reaction rate. In some cases the oxide formed is detrimental to further treatment, e.g. welding or dross removal, in which case the best results may be obtained with no oxygen (26) or just sufficient oxygen to aid absorption by the formation of an oxide layer.

3.5.4. Material Properties

3.5.4.1. Effect of Optical Properties - Reflectivity: For an opaque material the absorption = (1- reflectivity). Therefore one might expect that the high reflectivity materials would be more difficult to cut. This is the case, but not quite as dramatically as the above argument suggests because the reflectivity is not only a function of the material but also the surface shape, surface films (such as oxides), and surface plasmas. Due to the important effect of thin films such as oxides, the absorption can be strongly time dependent (31). Also due to the coupling effect with plasmas and the known decrease in reflectivity with temperature there is a further cause of a time dependency in the absorption as the material heats up. There is also a significant difference in the cutting rate depending on the surface finish as given in Table 3.6. Presumably wave formation on the melt front film would have a noticeable effect on the absorption but this has yet to be shown. Kielman (32) introduced the concept of "stimulated absorption" based upon the standing wave pattern on the cut front. The electromagnetic standing waves arise from reradiation from surface protruberances; a hard theory to visualise.

3.5.4.2. Effect of Thermal Properties: The ease with which a material can be successfully cut depends upon the absorptivity, the melting point of material or oxide formed, char tendency, and brittleness associated with the coefficient of thermal expansion. In fact the questions are:

1. Can sufficient power be absorbed?
2. Will this power cut successfully or damage the material?

Materials can be ranked by these properties, as in Table 3.7. See also the list of the cuttability of many materials in Industrial Laser Annual Handbook 1990 p3-6, published Penwell Books, Tulsa, Oklahoma, USA.

3.6. Examples of Applications

The main industrial application of lasers, at the present time, is in cutting. This work is increasingly done by laser "job shops". The costed example in Section 3.7 shows that the cost effectiveness of the laser is due to its speed and the high quality cut produced which reduces or eliminates after treatment and hence makes significant manufacturing cost savings. However these gains are only real if the laser can be kept working, due to the high capital investment involved in a laser facility (in 1990 around £150k-500k in equipment alone). Thus it makes sense to bring the work to the laser in the form of job shops. Some 90% of the present job shops offer a specialist service in cutting (33). There has been a considerable growth of job shops in recent years. Around 30% offer a service in laser engraving or marking using YAG or CO_2 lasers. The job shops are now being taken as a part of the manufacturing process and increasingly the job shop is being involved in the design stage of a component. This is partly as a result of new management techniques such as Just in Time (JIT) and MRPII and others. They are also increasingly being expected to take responsibility for the manufacture of complete components with a design team using the mutual expertise of the contractor and the job shop. This is a significant shift in manufacturing practice and has resulted in some remarkable cost effective developments. The applications span manufacturing industry from aerospace to food processing and toy manufacture.

Thus the applications for lasers in cutting are numerous and hard to list. It is the neatest and fastest profile cutting process. A typical job shop could have a turn around from drawing to article in a few hours - if pushed! Since the set up time is only that required to program the cutting table. This is hard to compete with unless the requirement is for more than 10,000 or so pieces when some hard automation, as with a stamping process would be cheaper.

Historically, laser drilling was the first industrial application by Western Electric, using a ruby laser in 1965 to drill holes in diamond dies for wire extrusion. A sample of some of these applications will give the flavour of where to look for further applications.

3.6.1. Die Board Cutting

One of the first industrial application of the laser. It used a BOC Falcon laser at 200W installed by William Thyne Ltd. in the UK in 1971. The die boards which are cut by this fully automatic machine are used in the manufacture of cartons. The laser replaced a process of sticking block board pieces together to make slots in which knives for cutting or creasing could be mounted. In the laser process the slots are simply cut in the block board and the knives mounted in the laser made groove. The process takes around 1/10th of the previous time. Nearly all cartons are now made this way with full CAD/CAM software to drive the laser and design the carton.

3.6.2. Cutting of Quartz Tubes

Quartz tubes are used for car halogen lamps. Thorn EMI uses 500W CO_2 lasers operating on a twin position cutting arrangement. The process was installed because there was a saving of material (approximately 1mm/cut at 4000 cuts/hr = 4m of tube/hr) and a significant reduction in fume and dust giving a saving in fume extraction and a better working environment.

3.6.3. Profile Cutting

This is mainly a job shop activity for the display industry, typewriter parts, gun parts, medical components, valve plates, gaskets and many others. Accuracy of cutting is around a few microns with a very fine finish for certain materials. One Australian firm specialises in making filter meshes, another in chain saw parts.

3.6.4. Cloth Cutting

Garment cutting is on the whole too slow by laser since the competing processes stack cut with a saw. Stack cutting cloth by laser is not easy due to welding, charring or smoke damage. Single thickness cutting of thick material is, however, excellent by laser. Thus it is used for cutting car floor carpets and seat covers. General Systems in Canada has a fully automatic laser machine using four lasers simultaneously to cut car fabrics.

3.6.5. Aerospace Materials

Hard and brittle ceramics such as SiN can be cut ten times faster by laser than by diamond saw.
Titanium alloys cut in an inert atmosphere are used in airframe manufacture. The laser saves around 17.6 man hours/plane for the Grumman Corp in the manufacture of one stabiliser component compared to chemical milling.
Aluminium alloys are similarly advantaged by using the laser, which has to be well tuned and of higher power. Savings of 60-70% costs compared to routing or blanking have been recorded.
Boron-epoxy and *Ti coated aluminium honeycomb* plates can also be cut by laser.
Stainless steel pressed parts are 3D profile cut by several aircraft manufacturers with a view to subsequent welding of the cut edge.

Fig. 3.23. Example of laser engraving.

3.6.6. Cutting Fibre Glass (34)

The advantages of cutting fibre glass by laser are the reduction of dust, no cracking of the edges and no tool wear, all of which are problems with drilling or sawing.

3.6.7 Cutting Kevlar

A nylon based epoxy armour plate used for a variety of reasons where strength and lightness are required has been one of the marvel materials of today. It is also a gift for the laser user since there are very few alternative techniques which can satisfactorily cut it (e.g. abrasive water jets). However, see Chapter 8 on safety, because the fumes can be poisonous.

3.6.8. Prototype Car Production (35)

The ease with which profiling can be done allows prototype production to be much faster compared to the use of nibblers, around ten times as many components can be made in a given time. The cutting of sun rooves in cars as an assembly option is now done by robotically guided lasers. Also the cutting of the holes for left or right hand drive vehicles is done by laser on the assembled car (36).

3.6.9. Cutting Alumina and Dielectric Boards

This is done both by through cutting or scribing. It is a common application.

3.6.10. Furniture Industry

Cutting timber of any hardness up to depths of 4cm is possible. Since there is no mechanical stress very tight nesting of parts can be arranged, giving significant saving of material. Cutting rates are, however, similar to those for a band saw. Charred surfaces, if not too badly charred, can be glued. Another application is laser engraving by machining very detailed patterns into wood, Fig. 3.23. This is done by rapidly scanning a focussed laser beam over a reflective mask, for example copper. Masks for drop out patterns are made by chemically etching thin copper sheets on a mylar backing. A development of this is to inlay the engraved area with metal or other wood as in marquetry. Some fine artistic work has been done this way, particularly for the Arab market.

3.6.11. Perforated Irrigation Pipes

A laser drilled hole does not have a burr flap and is less likely to block. Many dry areas use laser drilled irrigation pipes. A typical installation would use a 500W CO_2 laser drilling 4 holes/s of 0.5mm diameter in polythene pipe.

3.6.12. Perforating Cigarette Paper

By perforating cigarette paper the smoker is able to inhale air with the smoke; this reduces the nicotine content he breathes by condensation in the filter tip. This perforation is done at high speed (around 0.8m/s) making very neat, strong holes unlike the pin prick alternative. The productivity is generated by splitting the beam into four and scanning them through a mask onto the paper strip passing over a roll. Cutting paper by laser profile cutting has led to a new fashion in stationery and a near craze in 1989 in patterned sun tans!

3.6.13. Flexographic Print Rolls

Laser engraved rubber rolls are a precise and fast way of transferring a flat picture to a cylindrical roll for the printing of wallpaper and other articles.

3.6.14. Cutting Radioactive Materials

Work on radioactive materials is considerably easier with optical energy than other forms of energy since the generator is outside the hot zone and the only material which may become contaminated is the workpiece, fixtures and a few minor optics. It is also possible to transmit optical energy over long distances and so the work may not have to be confined to special areas. One of the advantages of the laser in cutting is the lack of fume. There is some fume but there is relatively little compared to any alternative. It is thus an attractive concept for the dismantling and repair of nuclear power stations.

3.6.15. Electronic Applications

Cutting of circuit boards has been mentioned. Resistance trimming of circuits, functional trimming of circuits and microlithography are new manufacturing processes introduced by the laser. The growing use of the excimer laser is of current interest. Hole drilling through circuit boards to join circuits mounted on both sides has advantages. The excimer can do this without risk of some form of conductive charring.

3.6.16. Hole Drilling

This is the normal area for YAG lasers but CO_2 lasers can also be used on many of these applications. Irrigation pipes and cigarette paper have been mentioned. Aerosol valve components, bubblers in gas liquid absorbers, holes in babies' teats, 0.04mm holes in spray nozzles for pumps, holes in turbine combustion chambers and blades, optical apertures and CDs and many others. The advantage of the laser is that it can drill holes at an angle to the surface; fine lock pin holes in monel metal bolts is an example. It is fast and accurate; For example in drilling Hastalloy - a nasty metal to drill because it is "gummy"; mechanical drilling is slow at around 60s/hole and causes extrusions at both ends of the hole which have to be cleaned; mechanical punching is fast but is limited to holes greater than 3mm diameter. ECM (Electro Chemical Machining) is too slow at 180s/hole but does give a neat hole. EDM (Electric Discharge Machining) is expensive and slow at 58s/hole; Electron beam drilling is fast at 0.125s/hole but needs a vacuum chamber and is more expensive than a YAG laser. A YAG laser took 4s/ hole. The holes were made by trepanning the required size over a range of sizes.

3.6.17. Scrap Recovery

Careful cutting of old telephone switches allows the recovery of the considerable precious metals content (37).

3.6.18. Laser Machining

This is similar to laser engraving on wood (Section 3.6.10); it has recently been achieved on steel (38). The rate of removal of material is slow being around 35mm³/min when using a 300W finely focussed beam and carefully designed nozzle operated at only 1 bar pressure.

3.7. Costed Example

Case 1

Shape to be cut in 1mm mild steel

42mm

80mm

	Press Tool	Laser
Capital Cost- Design and manufacture of Compound blank and die.	£1600	
Time/piece (for 300mm cut length @ 2.5m/min with a 500W laser).	0.5s	8.5s
Cost/piece (@£90/hr laser subcontract)	£1600/n	£0.22
Optimal production range - Breakeven	>7273	<7273
Delivery	6 weeks	1hr or so.

Case 2 Shape to be cut in 4mm stainless steel

48mm

60mm

	Press Tool	Laser
Capital cost in tooling	£5500	-
Time/piece for 600mm cut length @ 0.8m/min	1s	45s
Cost/piece (@ £90/hr subcontract laser)	£5500/n	£1.12
Optimal production range - Breakeven	>4910	<4910
Delivery	6-8 weeks	3hrs or so

The following case studies come from Laser Ecosse (39):

3.8.   Process Variations

3.8.1. Arc Augmented Laser Cutting

It has been found (40) that if an electric arc is located near the laser generated event then the arc will automatically root at the high temperature zone and the convective flow caused by this hot zone will constrict the arc to near the size of the laser beam for low current arcs up to 80A. Above this current the cathode jet from the arc is too strong and the two energy sources may not be located in the same place nor would the arc be constricted.   When the arc is on the same side of the workpiece as the laser then a damaged cut top edge will probably result. However when the arc is on the underside of the workpiece then the cutting process can be speeded up by a factor of around 2, as shown in Fig. 3.24 a plot of the cut speed against power.   Only where the curve starts to turn is the quality of the cut different from a pure laser cut. The weakness of this process is the need to have the arc on the opposite side to the laser. Nevertheless the process is there for anyone wishing to cut twice as fast for the same size of laser.

3.8.2. Hot Machining

The use of the laser to heat the material and thus soften it just prior to mechanical machining was the subject of considerable research in the USA during the early 80s. The process works by reducing the cutting force by about 50% provided that the heating is correctly located and does not allow for quenching from the bulk material in between heating and machining, thus transformation hardening instead of softening! The weakness with this process is the capital cost of the laser compared to a plasma torch which could be used to do

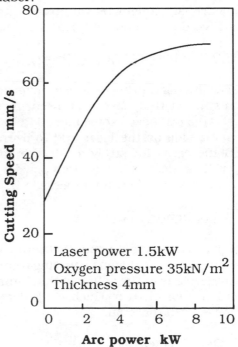

Fig. 3.24. Arc augmented laser cutting.

the same function.
3.9. Future Developments   (41)

The areas for development of laser cutting must be in the areas of:

1. Increasing the energy input to the cut region by having:
    a) Higher powered lasers
    b) Additional energy sources
    c) Improved coupling
    d) Smaller spot size

2. Increasing the ease of removal of the cut products - usually molten.
    a) Increase the drag
    b) Increase the fluidity

3.9.1. Higher Powered Lasers

Some significant results have been obtained with super pulsing to obtain instantaneous high powers. Too high power may create explosive conditions which spoil the quality. High powered lasers are being vigorously sought by various research programmes, notably the EU-REKA programmes in Europe.

3.9.2. Additional Energy Sources

Traditional cutting uses oxygen as an additional energy source and the results of that have been seen. Arc augmentation has been demonstrated but needs some invention to locate the arc successsfully on the same side as the laser and to overcome the arc initiation problems. Some areas for study are in the gas composition. One of the more original ideas is to blow iron powder into a cut in aluminium and thus start a small thermite process (14).

3.9.3. Improved Coupling

Polarisation and the coupling improvements with high power density (by increasing the plasma coupling route) are currently practised. Little is understood of the action of thin films on the melt front. There may be a certain gas mixture which would generate just the correct film thickness for a given material.

3.9.4. Smaller Spot Size

This would increase the power density and locate the energy more efficiently at the melt zone. It can be achieved by better designed laser cavities giving Gaussian mode beams and using shorter focal length lenses corrected for spherical abberration. It can also be achieved by using shorter wavelength lasers. The processing capabilities of a Gaussian high powered ultra violet beam has yet to be seen, but it should be awesome. However there is a safety aspect when using shorter wavelengths which will need to be addressed.

3.9.5 Increased Drag

This is the area of greatest promise. There are so many novel flow conditions which have not been examined yet. The whole subject of jet/ slot fluid flow is little understood at the time of writing.

3.9.6. Increased Fluidity

The melt could be made to flow better if the workpiece were vibrated ultrasonically, or the process were done under super gravity conditions, or the melt was more fluid because its chemistry is changed. For example to cut with chlorine instead of oxygen would give either a vapour halide and no dross or a very fluid dross. The chlorine gas is more dense than oxygen and therefore would have more drag. There is also an exothermic reaction involved. So chlorine should give improved performance - pity that the operator might get killed! In fact all these options need thinking about but seem difficult to engineer. Is it not true that the enjoyment of engineering is part dreaming, part doing?

3.10. Worked Example of Power Requirement

Question: It is hoped to cut 10mm thick mild steel by laser. What power laser is required and how fast will it cut?

1. From Table 3.4 we find that mild steel requires $16J/mm^2$ to sever. That is $P/Vt = 16J/mm^2$. If $t = 10mm$ then $P/V = 160J/mm$.

2. Penetration of the laser into mild steel is a function of the focusability of the beam. 1kW of a low order mode beam would penetrate this thickness, but a higher order mode beam would find it difficult. See Fig. 3.4.b showing speed vs thickness.

Thus a qualitative judgement is required to determine the laser power for the required penetration. Let's say 2kW. In that case the expected speed would be 12.5 mm/s.

The burning rate in oxygen is around 10-15mm/s (see Fig. 3.19, kerf width vs speed). This speed is a little too close to 12.5 mm/s. If the burning rate were to be the faster then a very poor cut quality would result. The burning rate can be controlled by the gas composition. Thus - depending upon economics - a 4kW laser would be preferred and the expected cutting rate would be <u>25 mm/s</u> using pure oxygen. Process economic evaluation can now proceed from there.

References

1. Powell.J., Wykes.C. "A comparison between CO_2 laser cutting and competitive techniques" Proc. 6th Int. Conf. on Lasers in Manufacture (LIM6) Birmingham, UK, ed W.M.Steen, May 1989 publ. IFS publ. Ltd. UK pp135-153.
2. Gabzdyl.J. Ph.D. Thesis, London University, 1989.
3. Andrews.J.G, Atthey.D.R. 1976 J.Phys D: Appl.Phys 9, 2181.
4. Yilbas.B.S. "The absorption of incident beams during laser drilling" Optics and Laser Tech Vol. 8, pp27-32, 1986.
5. Metals Handbook Desk edition Publ. ASM, Metals Park, Ohio 44073 1985.
6. Gagliano.F.P., Paek.U.C. 1971 IEEE J.Quantum Electron QE-7 No.6 paper 3.3 pp277.
7. Harryson.R., Herbertson.H. "Energy distribution during laser machining of glass" Proc. 5th Int. Conf. on Lasers in Manufacture (LIM5) Stuttgart, FRG. Sept 1988 ed. H.Hugel, publ. IFS (publ.) Ltd. UK pp25 - 34.
8. Laser cutting data
9. Powell.J. "Guidelines and data on laser cutting" Ind. Laser Annual Handbook 1990, publPenwell Books, Tulsa, Oklahoma, USA, 1990 pp56-62.
10. Duley.W.W., Gonsalves.J.N. 1972 Can.J.Phys. 50, 215.
11. Olsen.F.O., Emmel.A., Bergmann.H.W. "Contribution to oxygen assisted CO2 laser cutting" Proc. 6th Int. Conf. on Lasers in Manufacture (LIM6) Birmingham, UK, ed W.M.Steen, May 1989 publ. IFS publ. Ltd. UK pp67-79.
12. Perry's Chemical Engineers' Handbook publ McGraw Hill Publ Co London 6th edition 1984 Section 10.12.
13. Schouker.D. "The Physical Mechanism and Theory of Laser Cutting" Industrial Laser Annual Handbook, 1987 publ Penwell Books, Tulsa, Oklahoma, USA, pp65-79.
14. Powell.J. "CO_2 Laser Cutting" publ Carl Hanser Verlag, Munich FRG, 1990.
15. Duley.W.W. "Excimer Laser Etching of Organic Polymers" Proc LAMP Conf, Osaka, Japan, 1987, publ by High Temp Soc Japan 1987.paper 7A02 pp585-594.
16. Tonshoff.H.K., Butje.R. "Material processing with excimer lasers" Proc. 5th Int. Conf. on Lasers in Manufacture (LIM5) Stuttgart,

FRG. Sept 1988 ed. H.Hugel, publ. IFS publ. Ltd. UK pp35-47.

17. Bunting.K.A., Cornfield.G. 1975 Trans ASME J.Heat Trans (Feb) 116.

18. Sharp.C.M. "CO$_2$ laser cutting of highly reflective materials" Proc. 6th Int. Congress on Appl. of Lasers and Electrooptics ICALEO'87 San Diego, USA Nov 1987 publ Springer-Verlag/IFS 1988 pp149-153.

19. Shaw.L.H., Cox.M.J. "High Aspect Ratio Nd/YAG Laser Welding" Proc Conf Advances in Joining and Cutting Processes '89 paper 53, Harrogate Oct 1989 publ TWI, Abington, UK.

20. Olsen.F. 1981 Proc. Laser 81 Optoelectronics Conf, Munich publ. Springer, Berlin pp227-231.

21. Sato.S. et al. "Cutting of steels by high power CO laser beam" Proc 7th Int Conf on Applications of Lasers and Electrooptics (ICALEO'88) Oct, Santa Clara, ed G.Bruck, publ Springer Verlag/ IFS 1988 pp324-331.

22. Spalding.I.P. EUREKA EU 119 report, publ Culham labs, Abingdon, Oxon, UK 1990.

23. Steen.W.M., Kamalu.J.N. "Laser Cutting" Ch.2 in Laser Material Processing ed M.Bass publ. North Holland Publ Co., Amsterdam, Holland 1983.

24. Kamalu.J.N., Steen.W.M. 1981 TMS paper A81-38 publ AIME.

25. Fieret.J., Terry.M.J., Ward.B.A. "Aerodynamic interactions during laser cutting" Proc. SPIE conf. Laser Processing Fundamentals, Applications and Systems Eng. ed W.W.Duley, Quebec, Canada June 1986 Vol 668 pp53-62.

26. Weick.J.M., Bartel.W. "Laser cutting without oxygen and its benefits for cutting stainless steel" Proc. 6th Int. Conf. on Lasers in Manufacture (LIM6) Birmingham, UK, ed W.M.Steen, May 1989 publ. IFS publ. Ltd. UK pp81-89.

27. Kawasumi.H. "Laser processing in Japan" Ind. Laser Annual Handbook 1990, publPenwell Books, Tulsa, Oklahoma, USA, 1990 pp 141-143.

28. Gabzdyl.J.T., Steen.W.M., Cantello.M. "Nozzle beam alignment for laser cutting" Proc.ICALEO'87 San Diego USA May 1987, publ LIA Toledo 1988 pp143-148.

29. Zheng.H., Ph.D. Thesis London University 1990.

30. Forbes.N. 1976, Fabricator 6 No 5.

31. Duley.W.W. Optics and Laser Technology, 11,331, 1979.

32. Keilman.F. "Stimulated absorption of CO$_2$ laser light on metals" Proc NATO Advanced Studies Institute on Laser Surface Treatment, San Miniato, Italy Sept 1985 pp17-22.

33. T.Feeley "Job Shop Industry Matures" Industrail Laser Annual Handbook 1990, publ Penwell Books, Tulsa, Oklahoma, USA pp144-148.

34. Nuss.R., Muller.R., Geiger.M. "Laser cutting of RRIM - poly-

urethane components in comparison with other cutting techniques" Proc LIM-5 Conf Stuttgart, publ IFS publ 35-39 High St, Kempston Bedford, UK.1988 pp47-57.

35. Roessler.D.M. "New laser processing developments in the automotive industry" Ind. Laser Annual Handbook 1990, publ Penwell Books, Tulsa, Oklahoma, USA, 1990 pp109-127.
36. Hanike.L. "Laser Technology within the Volvo Car Corp" Proc LIM-5 Conf Stuttgart, publ IFS publ 35-39 High St, Kempston Bedford, UK.1988 pp97-118.
37. Gagliano.F. "The use of CO_2 lasers to dismantle scrap telephone switches for improved salvage of precious metal content" Proc 7th Int conf on Applications of Lasers and Electrooptics (ICALEO'88) Oct Santa Clara, ed G.Bruck, publ Springer Verlag/IFS 1988 pp379-386.
38. Laser Ecosse conference handout Welding Institute meeting, Coventry, 1987.
39. Ebert.G., Sutor.U. "Lasercaving offers new machining method" Industrial Laser Review Aug 1990 pp23-26.
40. Clarke.J., Steen.W.M. "Arc augmented Laser Cutting" Proc Laser 78 Conf London March 1978.
41. Steen.W.M. "Future developments" Ch.5 Applied Laser Tooling ed Soares.O.D.D., Perez Amor.M. publ. Nijhoff, Dordrecht, Netherlands. 1987.

"Darling I think you went through a red light!"

"Never mind the power, its the frequency tells us where we are"

Chapter 4

Laser Welding

"If you can't beat them join them"
anon

"By uniting we stand, by dividing we fall"
John Dickinson (1732-1808) *The Liberty Song Memoirs of the Historical
Society of Pennsylvania vol xiv.*

═══

4.1. Introduction

The focussed laser beam is one of the highest power density sources
available to industry today. It is similar in power density to an electron
beam. Together these two processes represent part of the new technology
of high energy density processing. Table 4.1 compares the power density
of various welding processes.

At these high power densities all materials will evaporate if the energy
can be absorbed. Thus, when welding in this way a hole is usually
formed by evaporation. This "hole" is then traversed through the
material with the molten walls sealing up behind it. The result is what
is known as a "keyhole " weld. This is characterised by its parallel sided
fusion zone and narrow width, Fig. 4.1. Since the weld is rarely wide
compared to the penetration it can be seen that the energy is being used
where it is needed in melting the interface to be joined and not most of
the surrounding area as well. A term to define this concept of efficiency
is known as the "joining efficiency". The joining efficiency is not a true
efficiency in that it has units of (mm² joined /kJ supplied). It is defined
as [Vt/P], the reciprocal of the specific energy, referred to in Chapter 3,

Table 4.1	Relative Power Densities of Different Welding Processes	
Process	Heat Source Intensity W/m^2	Fusion zone profile
Flux Shielded Arc Welding	$5 \times 10^6 - 10^8$	
Gas Shielded Arc Welding	$5 \times 10^6 - 10^8$	low / high
Plasma	$5 \times 10^6 - 10^{10}$	low / high
Laser or Electron Beam	$10^{10} - 10^{12}$	defocus / focus

Table 4.2	Relative Joining Efficiencies of Different Welding Processes
Process	Approximate Joining Efficiency mm2/kJ
Oxy Acetylene Flame	0.2 - 0.5
Manual Metal Arc (MMA)	2 - 3
Tungsten Inert Gas (TIG)	0.8 - 2
Submerged Arc Welding (SAW)	4 - 10
High Frequency Resistance Welding	65 - 100
Electron Beam (EB)	20 - 30
Laser	15 - 25

where V = traverse speed, mm/s;
t = thickness welded, mm; P =
incident power, kW. Table 4.2
gives some typical values of the
joining efficiency of various welding processes.

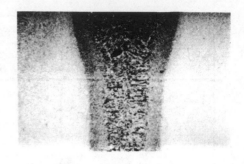

The higher the value of the joining
efficiency the less energy is spent
in unnecessary heating - that is
generating a HAZ or distortion.
Resistance welding is by far the
best in this respect because the
energy is only generated at the

Fig. 4.1. Micrograph of the transverse
section through a laser weld showing the
fusion and heat affected zones (HAZ).

high resistance interface to be welded. But it can be seen that the EB
and laser are again in a class by themselves. So how do they compare
with other processes in their performance characteristics and can they
be distinguished from each other? In fact what sort of market expectation can one foresee for laser welding? Is it a gimmick or a gift? The main
characteristics of the laser to bear in mind are listed in Table 4.3. The
ways in which these characteristics compare for alternative processes
are listed in Table 4.4.

It can be seen from these tables that the laser has something special to
offer as a high speed, high quality welding tool. However, Figs. 4.2 and
4.3 show that, at only 17% of applications for Nd-YAG and 14% for CO_2
lasers, it has been slow to penetrate the TIG market for high speed
welding processes, such as tube welding. This is mainly because of
uncertainty about the use and reliability of lasers compared to the TIG
process. The main market for laser welding processes is usually found
in areas requiring the welding of heat sensitive components such as
heart pace makers, pistons assembled with washers in situ, diaphragms
with sealed gas or electronic components. Another application area is in
welding magnetic or potentially magnetic material such as gears for
cars. The speed and neatness of the weld is, however, a challenge to the
future. Much research is currently being applied to weld cars, cans,
domestic equipment and aircraft. The process has many superior
qualities; so possibly we are only waiting for the market to be educated
in the use of lasers before it is more widely used.

4.2. Process Arrangement

As with laser cutting, welding relies on a finely focussed beam to achieve
the penetration. The only exception would be if the seam to be welded
is difficult to track or of variable gap; in which case a wider beam would

Table 4.3	Main Characteristics of Laser Welding
Characteristic	Comment
High energy density -"keyhole" type weld	Less distortion
High processing speed	Cost effective (if fully employed)
Rapid start/stop	Unlike arc processes
Welds at atmospheric pressure	Unlike EB welding
No X-rays generated	Unlike EB
No filler required (autogeneous weld)	No flux cleaning
Narrow weld	Less distortion
Relatively little Heat Affected Zone (HAZ)	Can weld near heat sensitive materials
Very accurate welding possible	Can weld thin to thick materials
Good weld bead profile	No clean up necessary
No beam wander in magnetic field	Unlike EB
Little or no contamination	Depends only on gas shrouding
Relatively little evaporation loss of volatile components	
Difficult materials can sometimes be welded	
Relatively easy to automate	General feature of laser processing
Lasers can be time shared	General feature of laser processing

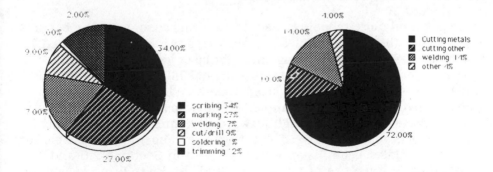

Fig. 4.2. Industrial applications of Nd-YAG lasers 1986

Fig. 4.3. Industrial applications of CO. lasers 1986.

Table 4.4	Comparison of Welding Processes				
Quality	Laser	EB	TIG	Resistance	Ultrasonic
Rate	✔	✔	✘	✔	✘
Low heat input	✔	✔	✘	✔	✔
Narrow HAZ	✔	✔	✘		✔
Weld bead appearance	✔	✔	✘		✔
Simple fixturing	✔	✘	✘		
Equipment reliability	✔		✔	✔	
Deep penetration	✘	✔		✘	
Welding in air	✔	✘		✔	
Weld magnetic materials	✔	✘	✔	✔	✔
Weld reflective material	✘	✔	✔	✔	✔
Weld heat sensitive mat.	✔	✔	✘	✘	✔
Joint access	✔			✘	✘
Environment, noise,fume	✔	✔	✘	✘	✘
Equipment costs	✘	✘	✔		
Operating costs	-	-	-	-	-

✔ point of merit; ✘ point of disadvantage.

be easier and more reliable to use. But, in this case, once the beam is defocussed the competition from plasma processes should then be considered. The general arrangement for laser welding is illustrated in Figs. 4.4 and 4.5. Fig. 4.5 illustrates the flexibility of the use of optical energy. It is in this area that laser users need to gain maturity. The advantages in welding, for example, a tube from the inside outwards, is that inspection becomes straightforward, and thus considerable quality control costs might be saved. The optical arrangements possible for focussing a laser beam have been discussed in Sections 2.6 and 2.7. Shrouding is a feature of all welding and the laser is no exception. However, shrouding is not difficult and coincides with the need to protect the optics from spatter. When welding high reflectivity material it is customary to tilt the workpiece by 5° or so, in order to avoid back reflections from entering the optics train and damaging 'O' rings or being reflected right back into the laser cavity and thus affecting the beam, the instant it is to be used. Such feed back has an air of lack of control and is a threat to the output window of the laser. It might, however, be a good thing if properly controlled in that one might expect greater power when the reflectivity is high.

4.3. Process Mechanisms - Keyholes and Plasmas

There are two modes of welding with the laser illustrated in Fig. 4.6. Conduction limited welding occurs when the power density is insuffi-cient to cause boiling - and therefore generate a keyhole - at the given

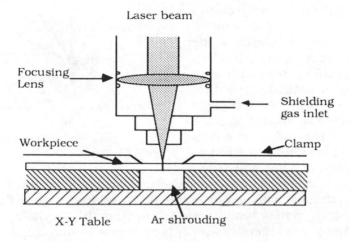

Fig. 4.4. General arrangement for laser welding.

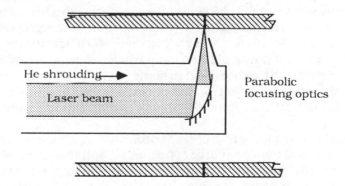

Fig. 4.5. Arrangement for welding pipe from the inside using metal optics.

welding speed. The weld pool has strong stirring forces driven by Marangoni type forces resulting from the variation in surface tension with temperature. This is discussed in more detail in the chapter on surface treatment, Chapter 6. Most surface treatments in which melting occurs, employ an out of focus beam which results in conduction limited weld beads. The alternative mode is "keyhole" welding in which there is sufficient energy/unit length to cause evaporation and hence a hole in the melt pool. This hole is stabilised by the pressure from the vapour being generated. In some high powered plasma welds there is an apparent hole, but this is mainly due to gas pressures from the plasma or cathode jet rather than from evaporation. The "keyhole" behaves like an optical black body in that the radiation enters the hole and is subject to multiple reflections before being able to es-

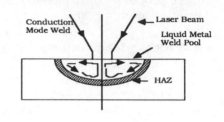

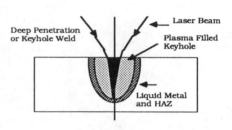

Fig. 4.6. Conduction limited and "keyhole" type welds.

cape. (Fig. 4.7). In consequence nearly all the beam is absorbed. This can be both a blessing and a nuisance when welding high reflectivity materials, since much power is needed to start the "keyhole" but as soon as it has started then the absorptivity jumps from 3% to 98% with possible damage to the weld structure.

Some ingenious experiments were done by the laser group at Osaka who photographed the "keyhole" during the laser welding of quartz and aluminium (1). In both it was seen that the "keyhole" has an approximate shape as shown in Fig. 4.7. and a flow pattern, illustrated in Fig. 4.8 (1). The flow pattern was followed by inserting high melting point particles and watching them with X-rays. The downward flow in this last example may be part gravitational since the particles were of tungsten. Nevertheless the twin vortices are a fact in some keyholes of sufficient depth. The location of the collision of the counter rotating vortices is a region susceptible to the trapping of bubbles and hence vulnerable to porosity.

Inside the keyhole there is considerable metal vapour, which is partially absorbing and hence capable of becoming hotter and forming a plasma. This hot plasma vapour emerging from the keyhole may ionise the shroud gas. Ionised gas has free electrons and is thus capable of absorbing or even blocking the beam. Fig. 4.9 shows what happens if

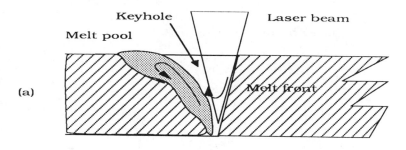

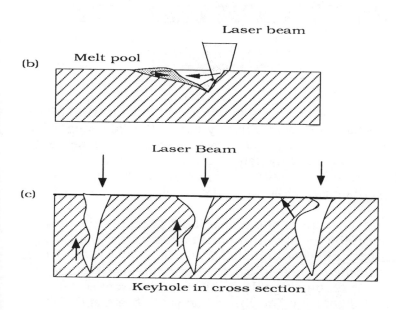

Fig. 4.7. Approximate shape and flow pattern in laser welds.

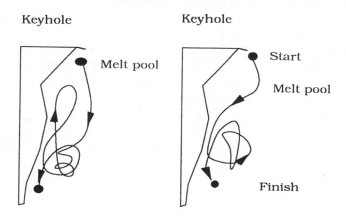

Fig. 4.8 Flow in a "keyhole" weld mapped by tungsten pellet (1)

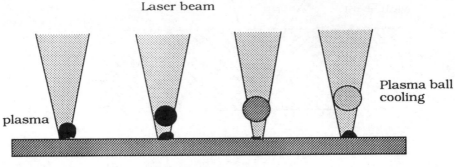

Fig. 49 Illustration of the blocking effect of the plasma if there is no side jet removing it.

there is no gas to blow the plasma away when welding with 10kW of laser power. The plasma forms intermittently due to the "blocking" of the beam. There is some discussion - aired in Chapter 2 - on whether the plasma is opaque enough at the temperatures measured to block the beam or whether the effect just noted is due to the plasma scattering the beam by variations in refractive index.

There are two principle areas of interest in the mechanism of keyhole welding. The first is the flow structure since this directly affects the wave formation on the weld pool and hence the final frozen weld bead geometry. This geometry is a measure of weld quality. The second is the mechanism for absorption within the keyhole which may affect both this flow and entrapped porosity. The absorption of the beam is by Fresnel absorption (absorption during reflection from a surface) and plasma reradiation. The Fresnel absorption can be calculated for a given shape of the leading edge of the keyhole to be non uniform (2). The calculation must allow for the slope of the face, mode structure of original incident beam, polarisation effects and focal position. The plasma effects vary with polarisation and speed - see Section 4.4.3. The Arata film (1) showed that ripples in the leading edge of the keyhole (Fig. 4.7) - on which nearly all the power first falls - may act as sites for explosive vaporisation sending a vapour cloud into the melt pool. Under certain conditions this may freeze as regular porosity at the root of the weld. There is a neck at the top of the keyhole which may again trap vapour.

4.4. Operating Characteristics

The main process parameters are illustrated in Fig. 4.10. They are:

Beam Properties:	Power, pulsed or continuous
	Spot size and mode
	Polarisation
	Wavelength
Transport Properties:	Speed
	Focal position
	Joint geometries
	Gap tolerance
Shroud Gas Properties:	Composition
	Shroud design
	Pressure/velocity
Material Properties:	Composition
	Surface condition

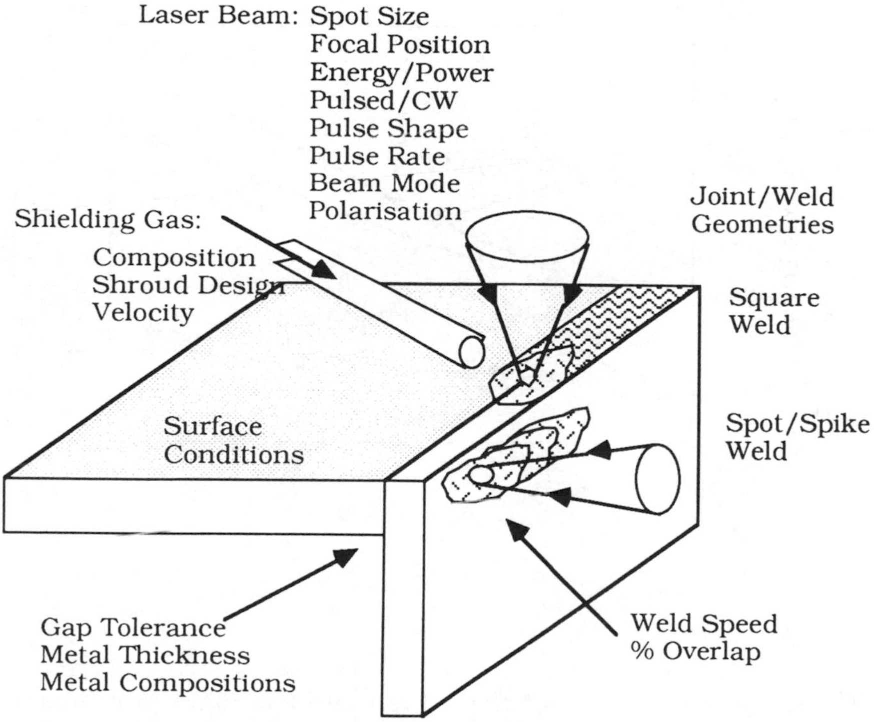

Laser Beam: Spot Size
Focal Position
Energy/Power
Pulsed/CW
Pulse Shape
Pulse Rate
Beam Mode
Polarisation

Shielding Gas:
Composition
Shroud Design
Velocity

Joint/Weld
Geometries

Square
Weld

Surface
Conditions

Spot/Spike
Weld

Gap Tolerance
Metal Thickness
Metal Compositions

Weld Speed
% Overlap

Fig. 4.10. The main process parameters.

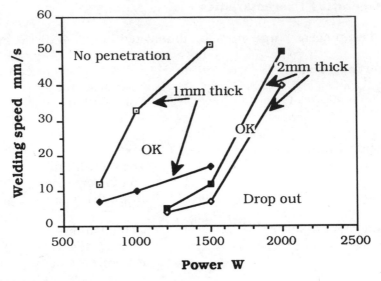

Fig. 4.11. Welding speed vs power for Ti-6Al-4V (3).

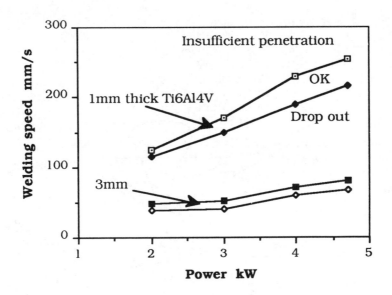

Fig. 4.12. Welding speed vs power for Laser Ecosse CL5 laser.

4.4.1. Power

4.4.1.1. Effect of Continuous Power: There are two main problems in welding: lack of penetration or the inverse, "drop out". These are the boundaries for a good weld for a given power as illustrated in Fig. 4.11 (3). The maximum welding speed for a given thickness rises with increase in power. The fall off shown at the higher power levels of 2kW is almost certainly due to the poorer mode structure given by most lasers when working at their peak power. However, for the results in Fig. 4.12(3), for higher power levels up to 5kW, the fall off may now be due to the same cause and also plasma effects. The main point to note from these two graphs is that for more power the operating window is larger. For high speeds the effects of sideways conduction during melting is slight and hence the Bessel functions discussed in the Swifthook and Gick model, Section 5.6, become soluble and an equation similar to that derived for cutting results. That is:

$$Y = 0.483 \ X \qquad\qquad\qquad (4.1.)$$

in which: $\qquad\qquad Y = 2vR/\alpha$ and $X = P/kgT$
$2R$ = weld width = w m $\qquad \alpha$ = thermal diffusivity = $k/\rho C_p$ m²/s
g = thickness $\qquad$ m $\qquad T$ = temperature, $\qquad\qquad$ K
P = power = $P(1-r_f)$ W $\qquad T_m.$ = melting point for width $\qquad$ K
r_f = reflectivity

Thus we have: $\qquad 0.483 \ P(1-r_f) = Vwg \ \rho C_p T_m$ $\qquad$ (4.2.)

This is a form of the lumped heat capacity model seen previously for cutting. This simple model for the maximum welding speed has neglected latent heat and so it must be a high value. It has also assumed that the power is distributed as a line source along the beam axis, the ultimate in fine focussing. However, the parametric relationships are enshrined in it. Incidentally this formula would act as a useful rule of thumb to find out what welding speed should be possible for a given laser power, if very finely focussed. It is usually high by around 30%.

Penetration is inversely proportional to the speed for a given mode, focal spot size and power, as shown in Fig. 4.13 (4).

4.4.1.2.   Pulsed Power: The use of pulsed power allows two more variables: pulse repetition frequency (PRF), and % overlap to be considered. The welding speed is decided by the spot size x PRF x (1-% overlap). In fact speed is independent of power. Penetration is a function of power and likewise the weld bead quality. Too much power causes vaporisation and material ejection as in drilling (5). Thus for welding the pulse is usually longer than for drilling and shaped to have a smaller initial peak.
<u>4.4.2. Spot Size and Mode</u>

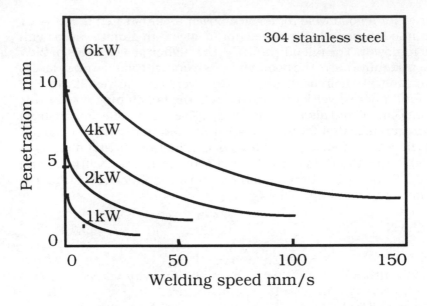

Fig. 4.13. Welding speed vs penetration for a fast axial flow CO2 laser (4).

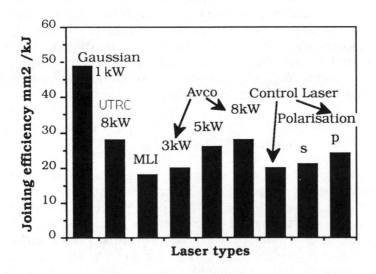

Fig. 4.14. Comparison of joining efficiencies for various lasers (6, BRITE1339).

The joining efficiency is greatly affected by the mode as illustrated in the results from Akhter (6), Fig. 4.14, on the welding of zinc coated steel using a variety of lasers. A similar study made by the Fraunhofer Institute at Aachen in comparing many lasers has shown that the superior mode structure of the Laser Ecosse AF5 laser gives the best penetration available today (7). This laser is fitted with flexible mirrors which allow accurate mode tuning to true TEM00 modes.

4.4.3. Polarisation

At first sight one might think that polarisation will have no effect on laser welding since the beam is absorbed inside a keyhole and hence it will be absorbed regardless of the plane of polarisation. Note this is quite unlike cutting where all the absorption had to take place on a steeply sloped cut front. This thought would be in essence correct but some second order events have been noted by Beyer et al. (2). Fig. 4.15 shows the slight variation in penetration thought to be due to polarisation effects. The resulting weld fusion zones are also wider for the case of s-polarisation (perpendicular to the plane of incidence) as expected since in this case the main absorption would be at the sides. The argument suggested for this phenomenon is that there are two absorption mechanisms. At slow speeds the plasma absorption dominates and the beam is absorbed by inverse Bremsstrahlung effects in the keyhole generating a plasma which appears blue in argon shrouded systems. As the speed increases, the Fresnel absorption (absorption by reflection on front face) gains in importance due to the cooler plasma being less absorbing. However, no polarisation effects were noted with aluminium. This is still a puzzle and throws some questions on the whole theory.

4.4.4. Wavelength

Due to the high absorptivity within the "keyhole" there is little operational difference when welding with long or short wavelengths. When welding with a conduction limited weld then the surface reflectivity becomes paramount and the lower reflectivity with the shorter wavelengths gives a distinct advantage to Excimer,YAG or CO lasers over the CO_2 laser.

4.4.5. Speed

The effect of speed on the welding process is principally described by the overall heat balance equation noted in Section 4.4.1.1. However in addition to these main effects there are some others. Firstly there is the

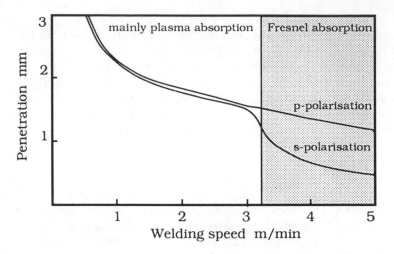

Fig 4.15 Influence of beam polarisation on welding performance (2).

effect of speed on the weld bead and secondly there is the problem of shrouding high speed welds.

4.4.5.1. Effect of Speed on the Weld Pool and Weld Bead Shape: As the speed increases so will the pool flow pattern and size change. In general the flow in a laser keyhole weld pool is shown in Figs. 4.6, 4.7 and 4.8. At slow speeds the pool is large and wide and may result in drop out, Fig. 4.16.d. In this case the ferrostatic head is too large for the surface tension to keep the pool in place and so it drops out of the weld leaving a hole or depression. This is described in detail by Matsunawa (8). At higher speeds, the strong flow towards the centre of the weld in the wake of the keyhole has no time to redistribute and is hence frozen as an undercut at the sides of the weld, diagrammatically shown in Fig. 4.16.b. If the power is high enough and the pool large enough then the same undercut proceeds and edge freezing occurs leaving a slight undercut but the thread of the pool in the centre has a pressure which is a function of the surface tension and the curvature (8). This leads to pressure instability causing the "pinch" effect in which those regions of high curvature flow to regions of lower curvature resulting in large humps, Fig. 4.16.c. The pressure, p, in these regions would vary by:

$$p = \gamma/r^2 \qquad\qquad\qquad (4.3.)$$

where γ = surface tension and r = radius of curvature.

There is an intermediate region in which there is a partial undercutting and central string. All this has been mapped for certain alloys by

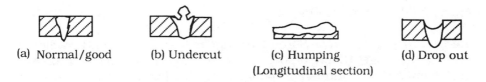

(a) Normal/good   (b) Undercut   (c) Humping   (d) Drop out
                                        (Longitudinal section)

Fig.4.16.Range of weld shapes varying usually with speed.

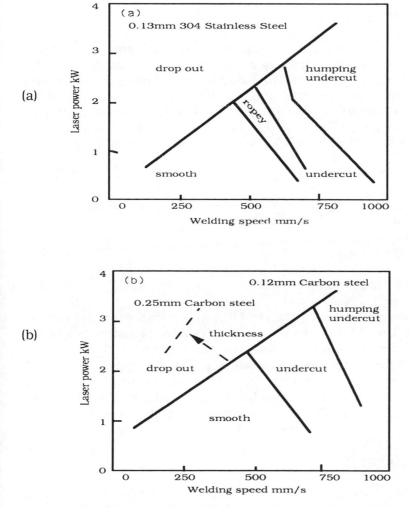

Fig. 4.17. Map of weld bead profiles as functions of welding speed and laser power
(9). a) 0.12 mm thick stainless steel. b) 0.12 and 0.25 mm thick mild steel.

Albright (9) as shown in Fig. 4.17.a,b.

4.4.5.2. Effect of Speed on Shroud Arrangements: The faster the welding process the longer the weld pool. A theoretical prediction of Gratzke et al (10) gives the relationship: pool length, $L = (3r^2 u)/2\alpha$ for a moving Gaussian source, but the results of marker experiments from Takeda (10) indicate the opposite. So there is room for discussion here! However with increase speed the hot metal extends further beyond the welding point. Thus trailing shrouds are usually needed to avoid atmospheric contamination.

4.4.6. Focal Position

There are suggestions (3,12,13) that the focal point should be located within the workpiece to a depth of around 1mm for maximum penetration. What one should consider here is the need to have sufficient power density to generate a "keyhole" and then for that power to stay together within the keyhole to increase the penetration. Thus the main parameters to consider would be the depth of focus and the minimum spot size. It has been shown in Section 2.6.2 that the depth of focus, z_f is given by:

$$z_f = \pm 15.7 \ F^2 \quad \mu m \text{ for } 10.6\mu m \text{ radiation} \tag{4.4}$$

and the minimum spot size, d_{min}, for a multi mode CO_2 beam, is given by:

$$d_{min} = 2.4(2p + 1 + 1)(F\lambda) \tag{4.5}$$

$$\therefore \ d_{min} = 80 \ F \quad \mu m \tag{4.6}$$

Fig. 4.18 shows the beam diameter vs distance from the lens for various F numbers. The shaded area shows the parabolic relationship between the depth of focus, z_f and the minimum beam diameter, d_{min}. A certain power density, P/d_{min} or P/d_{min}^2 is required to form a "keyhole" for a given traverse speed. This is marked in Fig. 4.18. by the horizontal line. From this analysis it can be seen that the optimal position of the focus for maximum penetration varies as shown in Fig. 4.19; a result in agreement with Seaman's work shown in Fig. 4.20.

4.4.7. Joint Geometries

4.4.7.1. Joint Arrangements: Laser beams causing keyhole type welds prefer a joint which helps the absorption and hence the formation of the keyhole. High intensity welding processes are not sensitive to different thicknesses of the pieces to be joined. This allows some new types of joint to be considered.

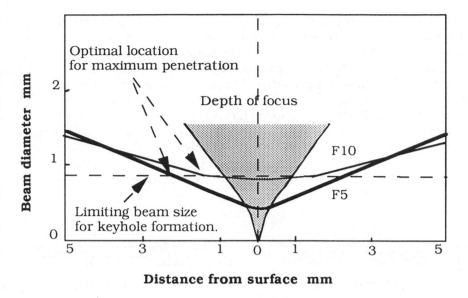

Fig. 4.18. Beam diameter vs distance from the focus.

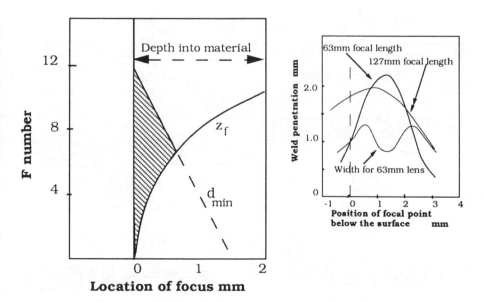

Fig. 4.19. Theoretical variation in the optimal position of the focus within the workpiece for maximum penetration, for a given power.

Fig. 4.20. Effect of focal position on weld penetration for 1018 steel (13).

Fig. 4.21 shows some of the variations which can be considered. The flare weld was used by Sepold (14) for very high speed welding of two strips at speeds up to 4m/s, and is currently used for making seam welds in thick section pipe. The plane of polarisation must be correct in this mode of welding or the beam will be absorbed before being reflected down to the point of the joint. It is, of course, a very efficient joining technique. The "Tee" weld geometry has a surprise attached to it, in that as the keyhole penetrates at an angle into the workpiece it tends to turn upwards to allow full penetration around the base of the tee. This very convenient event is the result of the reduced thermal load on the tee side of the keyhole, which encourages the melting isotherm that way.

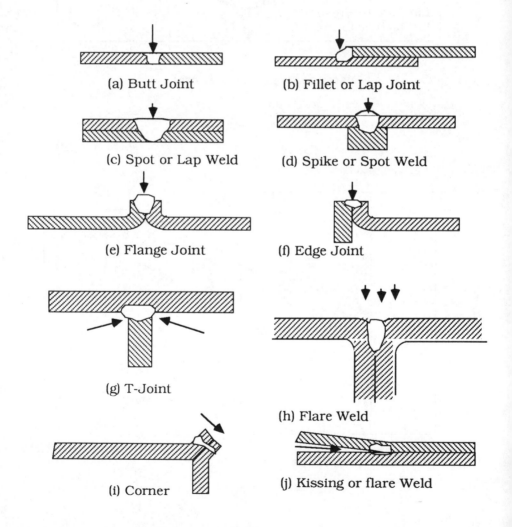

(a) Butt Joint

(b) Fillet or Lap Joint

(c) Spot or Lap Weld

(d) Spike or Spot Weld

(e) Flange Joint

(f) Edge Joint

(g) T-Joint

(h) Flare Weld

(i) Corner

(j) Kissing or flare Weld

Fig. 4.21 Various welding joint arrangements.

4.4.7.2. Effect of Gap: In butt joints the gap must be small enough that the beam can not pass straight through the joint. That is to suggest that the gap should be smaller than half the beam diameter (<200µm). For welds where there is a large gap the beam is sometimes rotated by rotating the lens off axis from the beam. However in these cases there is a chance of some drop out or a lower level in the weld. This can be corrected by adding filler material as a wire (15) or as a powder (6). On the whole the welds do not require filler material, they are "autogenous". One might question how this is possible when the conservation of mass suggests that if there is a gap there will be a fall in the level of the weld. In practice there is usually a rise in the level! This is due to the stresses in the cooling weldment drawing the workpieces together and so squashing the melt pool. Thus a small gap can be tolerated.

The extent of the squeeze is proportional to the forces which are in turn proportional to the contraction of the cooling weld. Thus the gap which can be tolerated, g, is approximately given by the relationship (6):

For butt welds: $A\beta\Delta Twt_p = gt_p$

$$\therefore \quad g = A\beta\Delta Tw \qquad\qquad (4.7)$$

where: β = Coefficient of thermal expansion.   m/°C
        ΔT = Temperature change, approx: melting point   °C
        w   = Weld width   m
        t_p = Sheet thickness   m
        g   = Gap width   m
        A   = Constant
        B   = Constant

For lap welds (gap between plates): $B\beta\Delta Tw\, 2t_p = gw$

$$\therefore \quad g = B\beta\Delta T\, 2t_p \qquad\qquad (4.8)$$

Welding with a gap in lap welding is essential if one is welding zinc coated steel or other material with a volatile coating. In this case there must be some way to vent the high pressure zinc vapour. Zinc boils at 906°C and steel melts at around 1500°C; the keyhole is even hotter. So as the keyhole enters the interlayer of zinc there is a sudden evolution of vapour which will destroy the weld continuity. Akhter(16) has calculated the required size of the gap from the volume of the zinc vapour to be exhausted. The situation, which is modelled , is shown in Fig. 4.22.a,b. The volume of vapour generated per second at the interface is:

$$2(w + 2b)Vt_{zn}\rho_s)/\rho_v \qquad m^3/s \qquad\qquad (4.9)$$

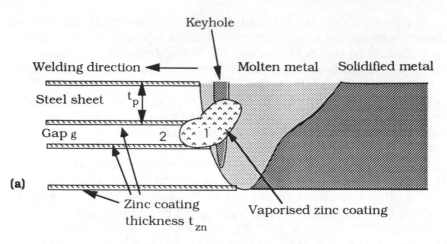

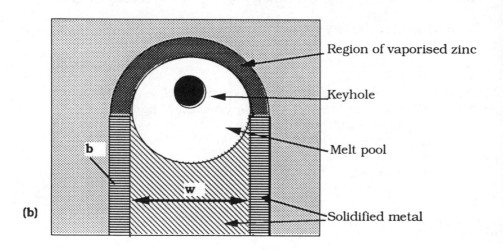

Fig. 4.22. Diagrams showing the welding of zinc coated steel with a small gap between the sheets for the exhausting of the high pressure zinc vapour generated during welding (15), (a) side view, (b) plan view.

The vapour escapes as it is formed around the melt pool at a velocity v_2. Thus the rate of escape of the vapour through the gap is:

$$= \frac{v_2 \pi (w + 2b) g}{2} \qquad (4.10.)$$

For a balance between the generation and exhaust of vapour we have from equations 4.9 and 4.10:

$$v_2 = \frac{4t_{zn} V \rho_s}{\pi \; \rho_v}$$
(4.11.)

This escape velocity can only be achieved with an acceleration pressure. This pressure must not exceed the ferrostatic head in the weld pool - $\rho_L g_c t_p$ - or the vapour will be expelled through the pool and destroy the weld quality. Thus:

$$v_2 = \sqrt{\frac{2\Delta P_{12}}{\rho_v}} = \sqrt{\frac{2\rho_L g t_p a}{\rho_v}}$$
(4.12.)

By eliminating v_2 between equations 4.11 and 4.12 we have a relationship for the limiting value of the gap required for sound welding of zinc coated steel:

$$g_{limit} = \frac{4t_{Zn} V \rho_s}{\pi \sqrt{2\rho_v \rho_L g t_p}}$$
(4.13.)

If the gap is smaller than this value then some blow out in the weld is to be expected. A method for controlling the gap in production has been suggested in (16) by dimpling. The model also suggests that there is a value of the laser power above which it is impossible to weld zinc coated steel. The value is expected to be around 5kW. This is the result of a balance between the exhaust of high pressure vapour, requiring a gap, and the mass balance on the weld pool to avoid drop out, not requiring a gap. The map of the process is shown in Fig. 4.23.

4.4.8. Gas Shroud and Gas Pressure

4.4.8.1. Shroud Composition: The gas shroud can affect the formation of plasma which may block the beam and thus the absorption of the beam into the workpiece.

The formation of plasma is thought to occur through the reaction of the hot metal vapours from the keyhole with the shroud gas. It is unlikely, in view of the fast emission of vapour from the keyhole that the shroud gas enters the keyhole. The plasma formed above the keyhole with the

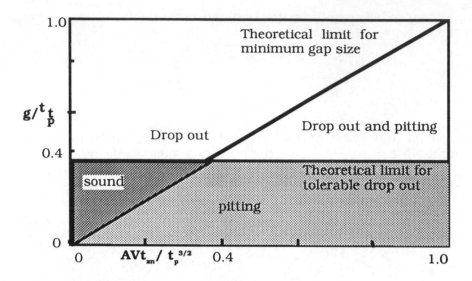

Fig. 4.23. Operational diagram for the welding of zinc coated mild steel with a gap

shroud gas will be absorbing to an extent determined by the temperature and the ionisation potential of the gases involved. Table 4.5 lists the ionisation potential of the gases often encountered in laser processing.

The plasma blocking effect will be less for those gases having a high ionisation potential. Thus helium is favoured, in spite of its price, as the top shroud gas in laser welding. The shroud underneath the weld would be of a cheaper gas, e.g. argon, N_2 or CO_2. The difference in penetration can be significant as shown in the results from RTM, Italy, Fig. 4.24. The plasma blocking is higher with higher powers. The results of Alexander (17) (Fig. 4.25) give this data a new slant. At slow speeds there is an advantage for helium but at high speeds there is an advantage for argon. The explanation is that the plasma is both good and bad in aiding absorption. If the plasma is near the workpiece surface or in the keyhole it is beneficial (as in Alexander's high speed results). If, however it is allowed to become thick or leave the surface its effect is to block or disperse the beam. This effect of speed on the preferred gas composition is also noted by Seaman (13). Because of this plasma effect it is usual to weld with a side blown jet to help blow the plasma away.

If the shroud gas is slightly reactive with the weld metal then a thin film of, say oxide , may form which will enhance the optical coupling. The work of Jørgensen (18) shows greater penetration when the shroud gas contained 10% oxygen, Fig. 4.26. This may, of course, be unacceptable for some welds, but is worth noting.

Table 4.5	Ionisation Potential of Common Gases and Metals (4,5)		
Material	1st Ionisation Potential, eV	Material	1st Ionisation Potential, eV
Helium	24.46	Aluminium	5.96
Argon	15.68	Chromium	6.74
Neon	15.54	Nickel	7.61
Carbon dioxide	14.41	Iron	7.83
Water vapour	12.56	Magnesium	7.61
Oxygen	12.50	Manganese	7.41

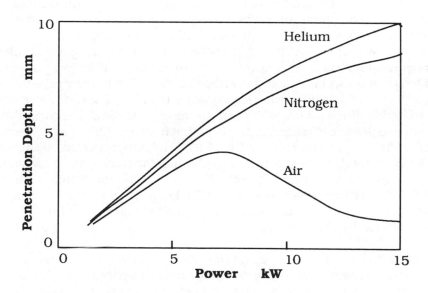

Fig. 4.24. Variation in penetration with shroud gas composition and laser power (results from RTM, BRITE 1339, 1991).

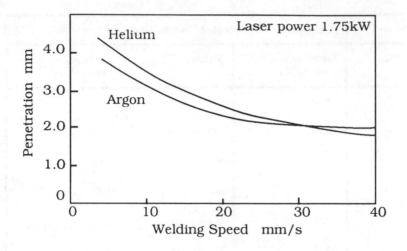

Fig. 4.25. Penetration vs speed for helium and argon shroud gases (17).

4.4.8.2. Effect of Shroud Design: The shroud design must give total coverage of the melt and the reactive hot region of the weld. It must do so without having flow rates which may cause waves on the weld pool. As just noted, in welding, a side jet is often added to blow the plasma away.   In the case of welding zinc coated steel, if the side jet may blow backward along the new weldment in which case the zinc vapour will condense on the weldment and so enhance the corrosion protection (19). The side jet can also be used to feed powder filler into the weld. For high speed welds the shroud will need to have a trailing section.   An interesting design, invented by the Welding Institute (20), was a plasma disruption jet. It is illustrated in Fig. 4.27. The concept is that if the fine 45° jet is correctly located it will blow the plasma back into the keyhole and hence enhance the absorption.   The welding performance is shown in Fig. 4.28. The main benefits are for thicker section welding. The weld fusion zone is altered to be more nearly parallel and the "nailhead" can be eliminated by this process.

4.4.8.3. Effect of Gas Pressure - as Both Shroud Velocity or Environment:
The nozzle pressure affects the gas flow rate and hence the ability of the gas to either blow the plasma away or correctly protect the weld. There is a minimum flow rate for adequate protection and also one for the removal of the plasma. There is a maximum rate above which the weld pool flow is affected and the melt pool ruffled causing a poor bead.

Variation in the environment pressure has a dramatic effect on penetra-

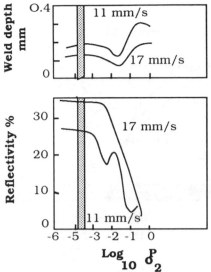

Fig 4.26. Weld depth as a function of partial pressure of oxygen, which is related to the reflectivity (after 18)

tion, particularly at very low pressures. The results are shown schematically in Fig. 4.29.(1). This means that the penetration of the laser and the electron beam are not too dissimilar. The EB is by necessity working in a high vacuum and hence enjoys the high penetration. There are two theories as to why this increase in penetration occurs. The first is that the lower pressure reduces the plasma density and hence the plasma is no longer blocking the beam. The second is that at the lower pressures the boiling point is reduced in a manner predicted by the Clapeyron-Clausius equation (21).

$$dp/dT = \Delta H/T\Delta V \qquad (4.14.)$$

Since the change in volume with the change in phase, ΔV, is negligible with melting, as opposed to boiling, there is not a similar effect on the melting point. Thus the melting point and boiling points become closer together at the lower pressures and hence the wall thickness of the keyhole will be thinner. A thinner liquid wall is easier to maintain and hence the keyhole is more easily stabilised. It is this stability which allows greater penetration. Arata (1) made a film illustrating the reduced plasma; it also shows the keyhole moving forward as illustrated in Fig. 4.29. To separate these two theories is very difficult and will test peoples' imagination for some time yet.

4.4.9. Effect of Material Properties

The main material problems with laser welding, in common with most welding methods are: crack sensitivity, porosity, HAZ embrittlement and poor absorption of the radiation. For welds of dissimilar metals there is the additional problem of the possible formation of brittle intermetallics.

Crack sensitivity refers to centreline cracking, hot cracking or liquation cracking. It is due to the shrinkage stress building up before the weld is fully solidified and strong enough to take the stress. It is thus most likely in metal alloys having a wide temperature range over which solidification occurs, e.g. those with high C, S, P contents. Some alloys listed in order of crack sensitivity, are given in Table 4.6. Cracking can

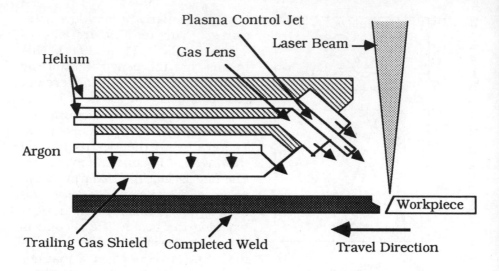

Fig.4.27. Illustration of the design of a plasma disruption jet with trailing shroud (20).

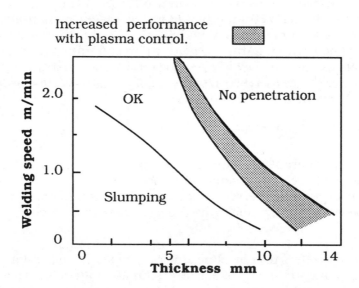

Fig. 4.28. The effect of a plasma control jet for a 6kW laser welding 18.8. stainless steel (20).

be reduced or eliminated by using a high pulse rate, adding a filler or using preheat.

Porosity often results when welding material subject to volatilisation such as brass, zinc coated steel or magnesium alloys. It may also be caused by a chemical reaction in the melt pool as with welding rimming steel or melting with inadequate shrouding such metals as, SG cast iron. It may also be present in metals having a high dissolved gas content such as some aluminiums. Control may be achieved with proper attention to the shrouding system, adding a "killing" agent such as aluminium to rimming steel or controlling the pulse rate or spot size.

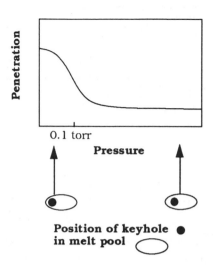

Fig. 4.29. Relationship between penetration and pressure for both electron beams and lasers.

The main advantages of laser welding are: that it is a process having a very low hydrogen potential (which may cause hydrogen embrittlement); it gives less tendency to liquation cracking due to the reduced time for segregation; and it causes less distortion due to the smaller pool size. Table 4.7 is a summary of some of the laser welding characteristics for the main alloy systems.

The welding of dissimilar metals is only possible for certain combinations as shown in Table 4.8. Due to the small fusion zone and relatively rapid solidification of the weld there is a greater range of welds possible with the laser than with slower processes. There is also a greater tendency to form metastable solid solutions as discussed in Chapter 6.

4.4.10. Gravity (22)

Duley's experiments in the NASA KC-135 microgravity aircraft using a 25W CW CO_2 laser to "weld" PMMA(acrylic), polypropylene, and polythene showed that for hypo and hyper gravity there was no change in the penetration depth, but there may be a reduction in the sheer strength with reduced gravity. In these experiments there was a notable

Table 4.6		Crack Sensitivity Rating of Certain Metals (4)								
Material	Crack tendency	Composition								
		C	Si	Mn	Cu	Fe	Ni	Cr	Mo	Other
Hastelloy B2	High	0.12	1.0	1.0		4-6	Rem	1.0	26	V,Co
Hastelloy C4		0.12	1.0	1.0		4.5-7	Rem	15	16	V,Co
Inconel 600		0.08	0.25	0.5	0.25	8.0	Rem	15.5	-	Al
Inconel 718		<0.08	-	-	0.15	18.5	52.5	19	3	Nb,Ti,Al
316 Stainless		0.08	1.5-3	2.0		Rem	19-22	23-26		
310 Stainless		0.25	1.5	2.0		Rem	19-22	24-26		
Hastelloy X		0.15				15.8	49	22	9	Co,W,Al
330 Stainless		0.08	0.7-1.5	2.0	1.0	Rem	34-37	17-20		
Aluminium 2024	Low			0.6	4.4	Mg 1.5	Al Rem			

Table 4.7	Laser Welding Characteristics for Different Alloy Systems
Alloy	Notes
Al Alloys	Problems with: 1. Reflectivity - requires at least 1kW 2. Porosity 3. Excessive fluidity - leads to drop out
Steels	O.K.
Heat Resistant Alloys: e.g. Inco 718, Jetehet M152, Hastelloy	O.K. but: 1.Weld is more brittle, 2. Segregation problems, 3. Cracking
Ti Alloys	Better than slower processes due to less grain growth
Iridium Alloys	Problem with hot cracking

change in the wave structure on the trailing edge of the keyhole. There were larger and faster waves with higher gravity.

4.5. Process Variations

4.5.1. Arc Augmented Laser Welding (23)

It has been found that the arc from a TIG torch mounted close to the laser beam interaction point will automatically lock onto the laser generated hot spot. Eboo (24) found that the temperature only had to be around 300 °C above the surrounding temperature for this to happen. The effect is either to stabilise an arc which is unstable due to its traverse speed or to reduce the resistance of an arc which is stable, Fig. 4.30. The locking only happens for arcs with a low current and therefore slow cathode jet; that is, for currents less than 80A. The beauty of this process is that the arc is on the same side of the workpiece as the laser.

Table 4.8 — Laser Weldability of Dissimilar Metal Combinations (4)

Legend: ■ = Excellent; G (shaded) = Good; F = Fair; P = Poor

	W	Ta	Mo	Cr	Co	Ti	Be	Fe	Pt	Ni	Pd	Cu	Au	Ag	Mg	Al	Zn	Cd	Pb	Sn
W																				
Ta																				
Mo																				
Cr		P																		
Co	F	P	F																	
Ti	F			G	F															
Be	P	P	P	P	F	P														
Fe	F	F	G				F													
Pt	■	F	G	G			F	P												
Ni	F	G	F	G			F	F	G											
Pd	F	G	G	G			F	F	G											
Cu	P	P	P	P	F	F	F	F												
Au	-	-	P	F	P	F	F	F												
Ag	P	P	P	P	P	F	P	P	F	P		F								
Mg	P	-	P	P	P	P	P	P	P	P	P	F	F	F						
Al	P	P	P	P	F	F	P	F	P	F	P	F	F	F	F					
Zn	P	-	P	P	F	P	P	F	P	F	F	G	F	G	P	F				
Cd	-	-	-	P	P	P	-	P	F	F	F	P	F	G	■	P	P			
Pb	P	-	P	P	P	P	-	P	P	P	P	P	P	P	P	P	P	P		
Sn	P	P	P	P	P	P	P	P	F	P	F	P	F	F	P	P	P	P	F	

The process allows a doubling of the welding speed for a modest increase in the capital cost. It does not enhance the penetration to any great extent (17). At very high speeds there may be some problems with the weld bead profile since the weld pool is larger than with the laser alone. The increased pool size is however not as much as expected. Eboo (23,24) showed by mathematical modelling the heat flow from the laser and the arc separately that the combined effect of the two was not the expected addition of two effects. The only way his results fitted the data was that the arc radius was decreased by the hot core from the laser event, Fig. 4.31. Thus arc augmented laser welding results in the arc rooting in the same location as the laser and doing so with a finer radius than usual - a true form of adding energy to the laser event. If augmenting the laser is not appealing, then the process has another advantage in that it is a method for guiding an arc.

4.5.2. Twin Beam Laser Welding

If two laser beams are used simultaneously then there is the possibility of controlling the weld pool geometry and hence the weld bead shape. Arata (1) using two electron beams demonstrated that the keyhole could be stabilised causing fewer waves on the weld pool and giving a better penetration and bead shape. O'Neill (25) using both an excimer and CO_2 beams simultaneously, showed that improved coupling for the welding of high reflectivity materials, such as aluminium or copper could be attained this way. The enhanced coupling was considered principally due to altering the reflectivity by surface rippling caused by the excimer blast (35MW for 20ns at 100Hz) with a secondary effect coming from coupling through the excimer generated plasma.

4.5.3. Walking Beams

Arata(1) suggested a method for avoiding the plasma by allowing the beam to dwell on a spot just long enough for the plasma to start forming and then to kick the beam forward to dwell on the next spot. He showed an improved penetration capability. The process is more efficient than pulsed welding.

4.6. Applications

The laser has a certain thickness range over which it competes as shown in Fig. 4.32. Within this range if the productivity of the laser can be used then it will usually compete successfully; as in the costed example

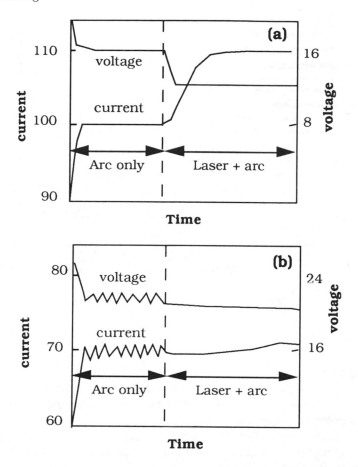

Fig. 4.30. The coupling of an arc and a laser beam results in the reduced resistance of the arc (a) or the stabilisation of the arc for high speed welding (b) (23,24).

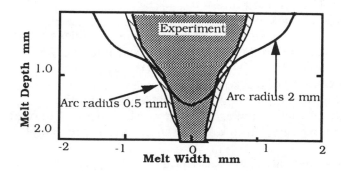

Fig. 4.31. Comparison of experiment with theory for various theoretical arc radii during arc augmented laser welding (from 24).

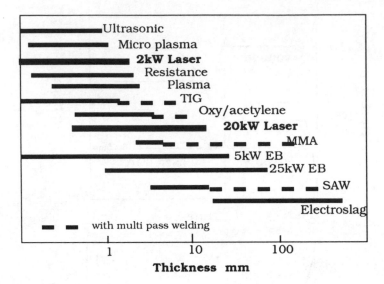

Fig. 4.32. Comparison of opereating range for different welding processes.

below.

Current applications are:
1. Welding of transmission systems for cars - taking advantage of the low distortion and the possibility of focussing near to potentially magnetic material.
2. Hermetically sealing electronic capsules.
3. Welding the end plate on a piston assembly for a car, while a nylon washer is nearby.
4. Welding transformer laminates to reduce hum- smaller weld zone reduces eddy losses.
5. Welding bimetallic saw blades.
6. Welding of stamped mufflers.
7. Welding of cooker tops.
8. Car doors, floor panels. Laser seam welding gives greater stiffness compared to spot welding and hence could lead to quieter cars.
9. Flare welding of thick pipes.
10. Welding of complex shapes prior to pressing, (26).
11. Repair of nuclear boiler tubes from the inside. There are nine internal welding units working in Japan at present (1991) (26).
12. Laser soldering is fast becoming a major process in the electronics industry.
The applications are too numerous to sensibly list them.   Roessler (27)

has given a useful review of applications in the car industry.

4.7. Costed Example

The cost of processing is made up of capital and operating elements. The relative capital cost of a laser facility is listed in Table 4.9.

If, for the sake of this example, we cost the capital as an operating cost at 10%/annum for 1800hrs/yr we can then compare the approximate cost of a metre weld in 3mm mild steel made by Manual Metal Arc (MMA) or laser as shown in Table 4.10.

The actual numbers are debatable but the difference is striking - it keys upon keeping the laser working. The capital depreciation is a fairly sensitive figure to the fraction of the year that the equipment is working. Therefore much of this work is done by job shops or in plant with a need for all year operation. A similar conclusion is drawn by the cost analysis of FIAT and COMAU (28).

As a general rule:
 "A laser working for one shift/day is just paying for itself; if working for two shifts/day, it is distinctly profitable; if it is working three shifts/day you will probably find the owner in the Bahamas or some such place". An

Table 4.9	Relative Capital Cost of a Laser Facility
Process	Capital Cost in Relative units (MMA = 1)
Manual Metal Arc(MMA)	1
Submerged Arc (SAW)	10
Electroslag	20+
TIG	2
Microplasma	20+
MIG	2
Resistance (Butt)	0.5-10
Oxy/Fuel	0.2
Electron Beam (EB)	10-450
Friction	4-100
Laser	100+

Table 4.10	Comparison of the Welding Costs for a Metre of Weld by Manual Metal Arc and Laser.	
	Manual Metal Arc (MMA)	Laser
Capital Cost	300A set --£850	2kW CO2 laser plus workstation £150,000.
Consumables	1m of 4mm flux coated mild steel rod	gases at £4/hr
Welding Speed	1mm/s	10mm/s
Process Time for 1m of weld	1000s	100s
	£	£
Capital depreciation at 10%/yr for 1800hrs/yr	0.013	0.23
Consumables: 1m of 4mm rod £4/hr gases	0.5	0.11
Labour £20/hr MMA; £15/hr Laser	5.50	0.41
Power at £ 0.06/kWh at 4kW MMA and 10kW laser	0.066	0.016
Clean up time at 40% arc time	2.20	-
Total	£8.28	0.77

interesting thought with which to finish this chapter!

References

1. Arata.Y. "Challenge of Laser Advanced Materials Processing" Proc Conf. Laser Advanced Material Processing LAMP'87, Osaka May 1987 publ High Temperature Soc. Japan pp3-11.(1987).
2. Beyer.E., Behler.K., Herziger.G. "Influence of Laser Beam Polarisation in Welding" Industrial Laser Annual Handbook 1990, publ Penwell Books, Tulsa, Oklahoma, USA, pp157-160 (1990).
3. Mazumder.J. "Laser Welding" Chapter 3 in Laser Material Processing, ed M.Bass publ North Holland Publ Co. Amsterdam, 1983 pp113-200.(1983).
4. Industrial Laser Annual Handbook 1990, publ Penwell Books, Tulsa, Oklahoma, USA, pp7-15 (1990).
5. Nonhof.C.J. "Material Processing with Nd-YAG Lasers" publ Electrochemical Publications Ltd., Ayr, Scotland (1988),pp192.
6. Akhter.R. Ph.D. Thesis, London University, 1990.
7. INPRO report LAS-11 Aug 1989 and LAS-12 Oct 1989 "Bestimmung der Strahlparameter und Schweisversuche an

Laserstrahlqellen mit Starhlleistungen P>5kW" INPRO, Nurnbergerstrasse 68/69, 1000 Berlin 30.

8. Matsunawa.A. "Role of surface tension in fusion welding" ptI J.W.R.I. (v11) No2 1982, ptII JWRI (v12) no.1 1983, ptIII JWRI v13) no.1 1984.
9. Albright.C.E., Chiang.S. "High Speed Laser Welding Discontinuities" Proc 7th Int.Conf. on Applications of Lasers and Electro Optics ICALEO'88 Oct/Nov 1988 Santa Clara Calif,USA. publ Springer Verlag/IFS.pp207-213 (1988).
10. Gratzke.U., Kapadia.P.D., Dowden.J. "Heat Conduction in HighSpeed Laser Welding" J.Phys D: App Phys. to be published 1991.
11. Takeda.T., Steen.W.M., West.D.R.F. "Laser Cladding with mixedpowder feed" Proc conf. ICALEO'84, Boston Nov 1984, pp151-158.
12. Wilgoss.R.A., Megaw.J.H.P.C., Clarke.J.N. "Assessing the laser for power plant welding" 1979 Weld Met Fabr. March 1979, 117.
13. Seaman.F.D. "Role of Shielding gas in High Power CO_2 CW Laser Welding" SME tech paper no.MR77-982 publ Society of Manufacturing Engineers, Dearborn, Mich, USA 1977.
14. Sepold.G., Rothe.R, Teske.K. "Laser Beam Pressure Welding - A New Technique" Proc Conf. Laser Advanced Material Processing LAMP'87, Osaka May 1987 publ High Temperature Soc. Japan pp151-156.(1987).
15. Norris.I.M. "High power laser welding of structural steels - current status" Proc. Conf. advances in Joining and Cutting Processes 89' Harrogate Oct 1989 publ The Welding Institute, Abington, Cambridge paper 55 1989.
16. Akhter.R., Steen.W.M. "The gap model for welding zinc coated steel sheet" Proc. conf Laser Systems applications in Industry Turin 7-9 Nov 1990 publ IATA 1991.
17. Alexander.J. Steen.W.M. "Effects of process variables on Arc Augmented Laser Welding" Proc Optica '80 Conf. Budapest, Hungary, Nov 1980.
18. Jørgensen.M 1980 Met Constr. Feb 1980 12 No (2), 88.
19. Akhter.R., Watkins.K.G., Steen.W.M. "Modification of the Composition of Laser Welds in Electrogalvanised steel and the effects on corrosion Properties" Journ Mat and Manf. Proc Vol 6 No.1(1991)
20. Oakley.P.J. "2 and 5kW Fast Axial Flow Carbon Dioxide Laser Material Processing" Proc. Material Proc Sym. ICALEO'82 Publ Laser Inst. America vol 31 pp121-128 1982.
21. Glasstone.S. Textbook of Physical Chemistry, publ Macmillan 1953 pp450.
22. Duley.W.W., Mueller.R.E. "Laser Penetration Welding in Low

Gravity Environment" Proc XXII ICHMT Int. Conf. on Manf. and Mat Proc. Dubrovnik Yugoslavia Aug 1990.

23. Steen W.M. "Arc Augmented Laser Processing of Materials" J. App Phys.   Nov. 1980 51 (11) pp5636-5641.

24. Steen W.M. , Eboo.M. "Arc Augmented Laser Welding" Metals Constr. VolII No.7 1979 pp332-336.

25. O'Neill.W., Steen.W.M. "Infra Red Absorption by Metallic Surfaces as a Result of Powerful u/v Pulses" Proc ICALEO '88 Santa Clara USA Nov 1988 publ Springer Verlag and LIA pp90-97.

26. Matsunawa.A. Proc SPIE conf ECO 4 1991 March 1991 Hague.

27. Roessler.D.M. "New Laser Processing Developments in the Automotive Industry" Industrial Laser Annual Handbook 1990 publ Penwell Books Tulsa, Oklahoma, USA pp109-127.

28. Marinoni.G., Maccagno.A., Rabino.E., "Technical and Economic Comparison of Laser Technology with the Conventional Technologies of Welding" Proc 6th Int Conf on Lasers in Manufacturing LIM6. Birmingham UK, May 1989 ed W.M.Steen, publ IFS publ Ltd. Bedford, UK pp105-120 (1989).

Chapter 5

Heat Flow Theory

"There is safety in numbers."
Jane Austin, Emma II,i

"We haven't the money so we've got to think."
Lord Rutherford 1871-1937 attr. in Prof R.V. Jones, 1962 Brunel
Lecture, 14th Feb 1962.

5.1. Introduction

When considering the prospect of mathematical modelling a process
there are at least two schools of thought. One says "why bother?" and
hopes it will go away; the other says "why not?" and plunges in with glee
regardless of direction. Both these attitudes miss the point that mathematical modelling is only a tool to help our understanding or control of
a process. The status of modelling has changed dramatically with easy
access to computers. Scientific reasoning no longer stops with the
derivation of a differential equation since these are now all soluble by
numerical methods. Software packages can be made which are the
models, and once made are available to anyone who wishes to turn them
on!

The target with modelling is:
1. Semi quantitative understanding of the process mechanisms
    for the design of experiments and display of results - dimen-
    sional analysis, order of magnitude calculations.

2. Parametric understanding for control purposes - empirical and
    statistical charts, analytic models.

3. Detailed understanding to analyse the precise process mechanisms for the purpose of prediction, process improvement and the pursuit of knowledge - analytic and numerical models.

The first two we have seen being used in the text of this book, and no doubt they are methods for which we have an instinctive appreciation. The dimensional analysis of heat flow problems implies a total knowledge of all the variables involved but no knowledge of how they are related. For any equation describing a physical relationship we know that the units must balance and that therefore there is a restriction on the number of ways the variables can be related. The way to take advantage of this is to arrange for the variables to be grouped as dimensionless groups, in which case the units will automatically balance, and we are left with a reduced number of variables - the groups. Once that is done then some experiments must be made to show how the groups are related. Buckingham's Π theorem (1) tells us how many groups to expect: the number of independent groups, $n = i-r$, where $i =$ number of variables and $r =$ the greatest number of these which will not form a dimensionless group (usually the same as the number of basic dimensions, except where there is some unusual symmetry). The usual dimensionless groups involved in heat transfer problems are:

The Fourier number,	$F = \alpha t/x^2$	a form of dimensionless time
The Peclet number,	$Pe = vD/\alpha$	a form of dimensionless velocity (ratio of convection to conduction)
The Reynolds number,	$Re = \rho vD/\mu$	another form of dimensionless velocity (ratio of viscous to inertial forces)
Dimensionless temperature,	$T^* = (T-T_o)/(T_1-T_o)$	
Dimensionless power,	$P/k\pi DT$	There are a few variations
Dimensionless distance,	x/D	

(α = thermal diffusivity; t = time; x= distance; v=velocity; ρ=density; μ=viscosity; P = incident absorbed power; k= thermal conductivity; D = characteristic distance, such as beam diameter; T = temperature).

Most experimental and theoretical results are likely to be reported using such groups because of the ecomony of expression achieved this way.

The third type of model involves complex mathematics or computing and is used as a means of inspiring awe or as a real tool of great importance. It is this type of model I wish to discuss in particular in this chapter. It should be remembered that the model is only a tool and as such has no significance other than in its use. Mathematical modelling is akin to astrology in that one can become mesmerised by the calculations and forget the underlying principles and objectives, the model has a sanctity of its own! (It can also become a "black hole" for research time.)

The questions we would like answered by a model and which are difficult or impossible to answer by experiment are:

1. How did the microstructure in a laser weld or surface treatment arise? We need to know thermal gradients, G; rates of solidification, R; time above certain temperatures; stresses; and stirring action or extent of mixing.

2. How did the weld bead form? There is a need to understand waves and movements in the weld pool and surface solidification processes.

3. What is the cause of cracking or distortion? We need to know the stress history, residual stresses.

4. What temperatures are likely in the environment of the laser event? We need to know depth of penetration of isotherms within thermally sensitive materials.

5. What is the optimum operating region? We need to be able to forecast experimental results, describe the effects of parameters not normally separable from others (for example the effect of the thermal conductivity is impossible to find by experiment since the density etc will also change if the material is changed) and forecast development opportunities.

This information can not be found from experiment, though the model results can be checked by experiment implying that the reasoning leading to the result is probably correct.

There are three major groups of models: the analytic models, numerical models and semi-quantitative models.

Since laser processing is usually a fairly quick process the heat flow by conduction is relatively confined and represents an approximately constant fraction of the delivered power. Due to this it is possible to use lumped heat capacity calculations for prediction purposes as seen in the severing and welding models. These have been derived in Sections 3. and 4. They constitute examples of the semi-quantitative models.

Nearly all other models in unsteady state heat transfer have to solve

Fourier's second law.   So this is our starting point.

(Fourier's First Law is $q = -kdT/dx$, where $q =$ heat flow/unit area. Baron Jean Baptiste Joseph Fourier (1768-1830) was a French mathematician famous for his representation of trigonometric functions as a series and his masterpiece "Theorie Analytique de la Chaleur" 1822 ,trans A. Freeman 1872 was considered one of the most important books of the 19th century. Through his association with Napoleon, he was governor of lower Egypt, and later Prefect of Isere where he wrote his work on heat. Though orphaned at 8 years old and of humble origins he was made a baron before he died.   His life in Egypt led him to believe that heat was healthy and so he spent much time swathed in blankets!)

Consider the differential element shown in Fig. 5.1. The heat balance on the element is:

Heat in - Heat out = Heat accumulated + Heat generated

The difference between the heat in and the heat out depends upon the different rates of conduction and convection.

Conduction:

Consider only the x direction (the other directions being analogous): Heat flow by conduction in the x direction is given by (heat in - heat out):

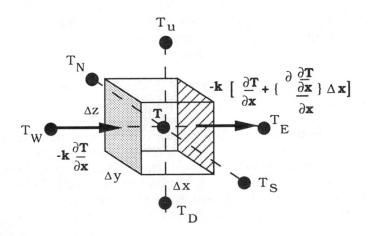

Fig. 5.1. Heat flow through a differential element.

$$= \left\{ -k\frac{\partial T}{\partial x} + k \left[\frac{\partial T}{\partial x} + \left\{ \frac{\partial \left(\frac{\partial T}{\partial x} \right)}{\partial x} \right\} \delta x \right] \right\} \delta y$$

$$= k \frac{\partial^2 T}{\partial x^2} \delta x \, \delta y \, \delta z$$

Total conduction in all three dimensions:

$$= k \left[\frac{\partial^2 T}{\partial x^2} + \frac{\partial^2 T}{\partial y^2} + \frac{\partial^2 T}{\partial z^2} \right] \delta x \, \delta y \, \delta z$$

which in vector notation is written: $= k \, \nabla^2 T \, \delta x \, \delta y \, \delta z$

Convection:

$$= \rho \, C_p \, U_x \, \delta z \, \delta y \, T - \rho \, C_p \, U_x \, \delta z \, \delta y \left(T + \frac{\partial T}{\partial x} \delta x \right)$$

$$= \rho \, C_p \, U_x \frac{\partial T}{\partial x} \delta x \, \delta y \, \delta z$$

which in vector notation is written: $= \rho \, C_p \, U_x \, \nabla T \, \delta x \, \delta y \, \delta z$

Accumulation: $\qquad\qquad = \rho \, C_p \frac{\partial T}{\partial t} \delta x \, \delta y \, \delta z$

Generation: $\qquad\qquad = H \, \delta x \, \delta y \, \delta z$

Total Balance: (divide throughout by $\delta x \, \delta y \, \delta z$)

$$k \, \nabla^2 T + \rho \, C_p \, U \, \nabla T - \rho \, C_p \frac{\partial T}{\partial t} = -H$$

$$\text{Thus} \quad \frac{U}{\alpha} \nabla T + \nabla^2 T - \frac{1}{\alpha} \frac{\partial T}{\partial t} = -\frac{H}{k} \qquad\qquad (5.1.)$$

This is the basic equation to be solved. (Meanings of symbols are listed at end of chapter.)

5.2. Analytic Models in One Dimensional Heat Flow

If the heat flows in only one direction and there is no convection or heat generation the basic equation becomes:

$$\frac{\partial^2 T}{\partial z^2} = \frac{1}{\alpha}\frac{\partial T}{\partial t}$$

(5.2.)

If it is assumed that there is a constant extended surface heat input and constant thermal properties, with no radiant heat loss or melting then the boundary conditions are:

At z = 0; Surface power density, $F_o = \{\frac{P_{tot}(1 - r_f)}{A}\} = -k\left[\frac{\partial T}{\partial z}\right]_{surf}$

At z = ∞; $\frac{\partial T}{\partial z} = 0$

At t = 0; $T = T_0$

The solution is:

$$T_{z,t} = \frac{2F_0}{k}\left\{(\alpha t)^{\frac{1}{2}}\, ierfc\left(\frac{z}{2\sqrt{\alpha t}}\right)\right\}$$

(5.3.)

Where the "integral of the complimentary error function", ierfc, means:

$$ierfc(u) = \frac{e^{-u^2}}{\sqrt{\pi}} - u\,[1 - erf(u)]$$

and since the error function, erf(u), is not freely available, as with logarithms, and it is difficult to maniplulate algebraically it can be substituted by a polynomial with an accuracy of one part in 2.5×10^5, which is usually sufficient for our needs. The polynomial is (2):

$$erf(u) = 1 - (a_1 b + a_2 b^2 + a_3 b^3)\,e^{-u^2}$$

where: b $= (1 + cu)^{-1}$ a_1 = 0.3480242
                                    a_2 = -0.0958798
                                    a_3 = 0.7478556
                                    c   = 0.47047

(Incidentally the error function erf(x) is defined as:

$$\text{erf}(x) = \left(\frac{2}{\sqrt{\pi}}\right)\int_0^x e^{-\xi^2}\, d\xi$$

This leads to the differential of the error function
derf(x)/dx = (2/√π) e^{x2}
and the values at extremes are: erf(o) = 0; erf(∞) = 1)

If the power is turned off then the material will cool according to the
relationship:

$$T_{z,t} = \frac{2F_0}{k}\sqrt{\alpha}\left[\sqrt{t}\ \text{ierfc}\left(\frac{z}{2\sqrt{\alpha t}}\right) - \sqrt{t_0 - t_1}\ \text{ierfc}\left(\frac{z}{2\sqrt{\alpha(t-t_1)}}\right)\right] \qquad (5.4.)$$

where the variables are:

T	= Temperature	C	k	= Thermal conductivity	W/mK
z	= Depth	m	α	= Thermal diffusivity	m²/s
t	= Time	s	t_0	= Time at start of power on	s
t_1	= Time power off	s	F_0	= Absorbed power density	W/m²

This model gives a feel for the extent of the conduction process and
applies when the flow is in one direction. That will be reasonably correct
when the heat source is large compared to the depth considered or that
the flow is in a rod or some such geometry. But this calculation has not
allowed any variation or concept of beam size or mode structure; speed
is only simulated by time on; and no allowance is made for workpiece
thickness.   Some typical solutions from(3) are given in Table 5.1.

The solutions from this model are useful.   Brienan and Kear (3)
developed graphs of the cooling rate, thermal gradient and solidification
rates expected with this form of heating they are shown for pure nickel
in Figs. 5.2-5.5.

Table 5.1.	Heating and cooling times as a function of power for a 0.025mm melt depth in nickel.		
Power Density W/cm2	Surface Temperature °C	Melting Time s	Cooling Time s
550 000	2730	6.71 x 10-5	3.02 x 10-5
200 000	1960	2.64 x 10-4	2.47 x 10-5
50 000	1590	2.77 x 10-3	2.24 x 10-5
5000	1469	0.236	2.17 x 10-5
500	1456	23.1	2.10 x 10-5

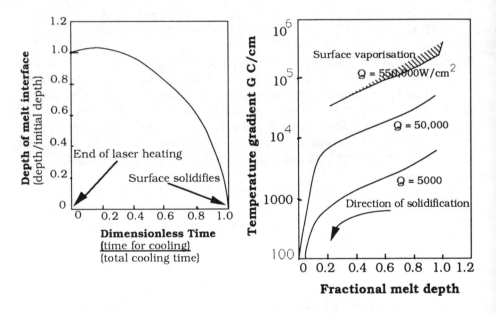

Fig. 5.2. Melt depth history (3) for pure nickel with 500,000 W/cm² absorbed power initial melt depth 1.2mm. Maximum temperature 2038°C.

Fig. 5.3. Transient behaviour of the temperature gradient at the melt front for pure Ni; initial melt depth 0.025mm.

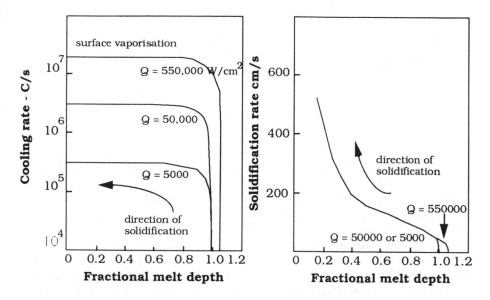

Fig. 5.4. Transient behaviour of the cooling rate. Conditions as for Fig. 5.3.

Fig. 5.5. Transient behaviour of the solidification rate. Conditions as for Fig. 5.3.

5.3. Analytic Models for a Stationary Point Source (4)

5.3.1. The Instantaneous Point Source

The differential equation for the conduction of heat in a stationary medium is:

$$\left[\frac{\partial^2 T}{\partial x^2} + \frac{\partial^2 T}{\partial y^2} + \frac{\partial^2 T}{\partial z^2}\right] = \frac{1}{\alpha}\frac{\partial T}{\partial t}$$

This is satisfied by:

$$T = \frac{Q}{8\,(\pi\alpha t)^{2/3}}\, e^{-\{(x-x')^2 + (y-y')^2 + (z-z')^2\}/4\alpha t} \tag{5.5.}$$

As $t \to 0$ this expression tends to zero at all points except (x',y',z') where it becomes infinite. Also the total quantity of heat in the infinite region is:

$$= Q\,\rho\,C \qquad \text{Which is as it should be.}$$

$$\int\limits_{-\infty}^{+\infty}\int\limits_{-\infty}^{+\infty}\int\limits_{-\infty}^{+\infty} \rho CT\, dx\, dy\, dz = \frac{Q\rho C}{8(\pi\alpha t)^{2/3}} \int\limits_{-\infty}^{+\infty} e^{\frac{-(x-x')^2}{4\alpha t}}\, dx \int\limits_{-\infty}^{+\infty} e^{\frac{-(y-y')^2}{4\alpha t}}\, dy \int\limits_{-\infty}^{+\infty} e^{\frac{-(z-z')^2}{4\alpha t}}\, dz$$

$$= Q\rho C \qquad \text{which is as it should be.}$$

Thus the solution above in equation 5.2 may be interpreted as the temperature in an infinite solid due to a quantity of heat $Q\,\rho C$ instantaneously generated at $t = 0$ at the point (x',y',z'). Q is seen to be the temperature to which a unit volume of the material would be raised by the instantaneous point source.

5.3.2. The Continuous Point Source

Since heat is not a vector quantity the effects from different heat sources can be added. Thus if heat is liberated at the rate of $\phi(t)\,\rho C$ per unit time from $t=0$ to $t=t$ at the point (x',y',z'), the temperature at (x,y,z) at time t is found by integrating equation 5.2 over that time period:

$$T(x,y,z,t) = \frac{1}{(8\pi\alpha)^{3/2}} \int\limits_{0}^{t} \phi(t')\, e^{\frac{-r^2}{4\alpha(t-t')}} \frac{dt'}{(t-t')^{3/2}}$$

where $r^2 = (x-x')^2 + (y-y')^2 + (z-z')^2$

If $\phi(t)$ is constant and equal to q, we have:

$$T = \frac{q}{4(\pi\alpha)^{3/2}} \int\limits_{1/\sqrt{t}}^{\infty} e^{\frac{-r^2\tau^2}{4\alpha}}\, d\tau \qquad\qquad \text{where } \tau = (t-t')^{-1/2}$$

$$= \frac{q}{4\pi\alpha r}\, \text{erfc}\left\{ \frac{r}{\sqrt{4\alpha t}} \right\} \qquad\qquad\qquad (5.6.)$$

As $t \to \infty$ this reduces to $T = q/4\pi\alpha r$, a steady temperature distribution in which a constant supply of heat is continually introduced at (x',y',z') and spreads outward into an infinite solid.

5.3.3. Sources Other than Point Sources

By integrating this point source solution over an area it is possible to calculate the heating from line sources, disc sources or Gaussian sources or any other definable distribution. Carslaw and Jaegar's book "Conduction of Heat in Solids"(4) is a collection of solutions for nearly every conceivable geometry.

For example the solution for the surface central point under a stationary Gaussian source can be shown to be:

$$T_{cont\,Gauss}(0,0,t) = \frac{2P(1-r_f)D}{\pi D^2 k \sqrt{T}} \tan^{-1}\left[\frac{2(\sqrt{\alpha t})}{D}\right] \tag{5.7.}$$

From which the equilibrium temperature of that spot would be:

$$T(0,0,\infty) = \frac{P(1-r_f)D}{D^2 k \sqrt{\pi}}$$

giving,

$$\frac{T\pi k D}{P(1-r_f)} = \sqrt{\pi} = 1.77 \tag{5.8.}$$

This is an interesting rule of thumb for calculating the maximum possible temperature, if the only loss mechanism is conduction.

5.4. Analytic Models for a Moving Point Source (5)

By integrating the point source solution over time and moving it by making $x = (x_o + vt)$ Rosenthal (5) developed the well known fundamental welding equations. He actually had three equations: The one dimensional solution for a melting welding rod; the two dimensional solution for a moving line source (simulating a keyhole on thin plate welding); and the three dimensional moving point source (simulating laser surface treatments or thick plate welding).

The solution for the moving point source assumes a semi-infinite work piece, no radiant loss, no melting and constant thermal properties over the temperature range concerned. The solution is:

$$T - T_0 = \frac{Q}{2\pi k} e^{-\frac{vx}{2\alpha}} \cdot \frac{e^{-\frac{vR}{2\alpha}}}{R} \tag{5.9.}$$

For the boundary conditions:

$$\frac{\partial T}{\partial x} \to 0 \text{ for } x \to \infty; \quad \frac{\partial T}{\partial y} \to 0 \text{ for } y \to \infty; \quad \frac{\partial T}{\partial z} \to 0 \text{ for } z \to \infty$$

and $\quad \frac{\partial T}{\partial R} 4\pi R^2 k \to Q \text{ for } R \to 0$

Where $R = \sqrt{x^2 + y^2 + z^2}$

From this the rate of cooling for the centreline surface spot ($y = 0$, $x > 0$, $z = 0$) can be derived as:

$$\frac{\partial T}{\partial t} = -2\pi k \left[\frac{v}{Q}\right] (T - T_0)^2 \qquad (5.10.)$$

From this it is possible to estimate the expected quench rates for various surface treatments, as shown in Fig. 5.6 using a 2kW laser on steel, k = 52W/mK; melting point 1500°C. These are fast cooling rates; faster rates can be achieved by pulsing. A 1ps pulse would give a quench rate of around 10^{13} K/s.

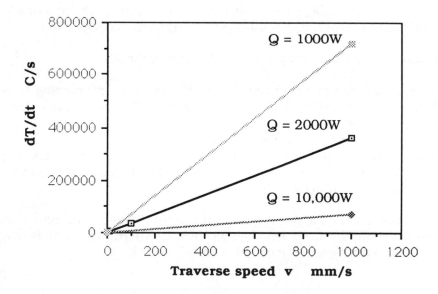

Fig. 5.6. Estimated cooling rates from equation 5.10.

5.5. Alternative Surface Heating Models

5.5.1. Ashby Shercliffe Model: - Moving Hypersurface Line Source (6)
There are many models published. They are mostly based on the
fundamental solution for a point source or line source. A significant one
for laser processing is that of Ashby and Shercliff(6).They derived a
solution, as a development of the Ashby-Easterling hyper point source(7),
in which the heat source is a moving finite line source situated above the
surface and parallel to it. The advantage of this is that it allows a pseudo
beam diameter to be considered. The assumptions are: the heat source
is a line source of finite width in the y direction and infinitesimally small
in the x or welding direction. The workpiece is homogeneous and
isotropic having constant thermal properties. No account is taken of
latent heat effects, radiation or surface convection heat losses. The
workpiece is semi infinite.

The solution for the temperature variation with time is:

$$T - T_0 = \frac{Aq/v}{2\pi\lambda \left[t \left(t + t_0 \right) \right]^{1/2}} \exp\left[- \frac{1}{4\alpha} \left\{ \frac{\left(z + z_0 \right)^2}{t} + \frac{y^2}{t + t_0} \right\} \right] \qquad (5.11.)$$

Cooling rates:

$$\frac{dT}{dt} = \frac{T - T_0}{t} \left\{ \frac{\left(z + z_0 \right)^2}{4\alpha t} - \frac{1}{2} \left(\frac{2t + t_0}{t + t_0} \right) \right\} \qquad (5.12.)$$

All this can be plotted in their "master plot" as shown in Fig. 5.7. This
model is convenient to use and fits transformation hardening problems
quite well.

5.5.2. Davis et al Model - Moving Gaussian Source (8)

This group solved the moving Gaussian surface heat source problem
and achieved a complex equation - as the reader might expect.
However they then derived an expression for the depth of hardness
assuming constant thermal properties and no surface heat losses by
convection or radiation;   they also assumed that there were no
latent heat effects.

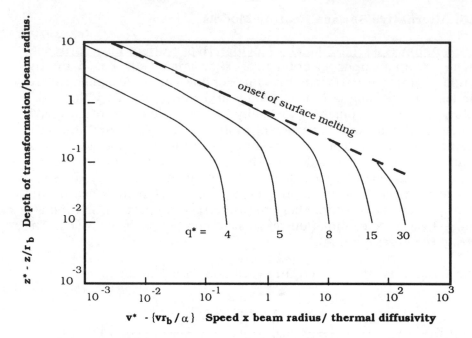

Fig. 5.7. Ashby/Shercliff "Master Plot" for predicting laser transformation depth of hardening (6). $q^* = AP/r_t\, k(T_{trans} - T_o))$ a dimensionless beam energy.

Their solution for the depth of hardness, d, is:

$$d = 0.76\, D \left[\cfrac{1}{\dfrac{\alpha' C}{q} + \pi^{1/4}\left(\dfrac{\alpha' C}{q}\right)^{1/2}} - \cfrac{1}{\dfrac{\alpha' C}{q_{min}} + \pi^{1/4}\left(\dfrac{\alpha' C}{q_{min}}\right)^{1/2}} \left(\dfrac{q_{min}}{q}\right)^{3} \right] \qquad (5.13.)$$

Where $\alpha' = \dfrac{\rho u D C}{2k}$; $q = \left[\dfrac{2P(1 - r_f)}{D\pi^{3/2}k\,(T_c - T_0)}\right]$; $q_{min} = (0.40528 + 0.21586\,\alpha\, C)^{1/2}$.

where q_{min} is the minimum absorbed power for which any hardening can occur and

$$C = C_\infty - 0.4646\,(C_\infty - C_0)\,(\alpha\, C_0)^{1/2}$$

The constants vary with the material. For En8 they are: $C_\infty = 1.599$, $C_o = 0.503$.

Although this expression may look at first sight to be a bit formidable, it is simple to solve on a personal computer. The results have been shown to fit experiments very well (9).

5.6. Analytic Keyhole Models - Line Source Solution

The Rosenthal solution for the moving line source (5) assumes that the energy is absorbed uniformly along a line in the depth direction. The assumptions are as for the other Rosenthal models. Fig. 5.8 illustrates the heat flow pattern. The solution simulates single pass welding in thin materials or a fully penetrating keyhole weld. It is (5):

$$T - T_0 = \frac{Q}{2 \pi k g} e^{+\frac{vx}{2\alpha}} K_0 \left[\frac{vR}{2\alpha} \right] \qquad (5.14.)$$

The boundary conditions are:

$$\frac{\partial T}{\partial x} = 0 \text{ as } x \to \infty; \quad \frac{\partial T}{\partial y} = 0 \text{ as } y \to \infty; \quad \frac{\partial T}{\partial z} = 0$$

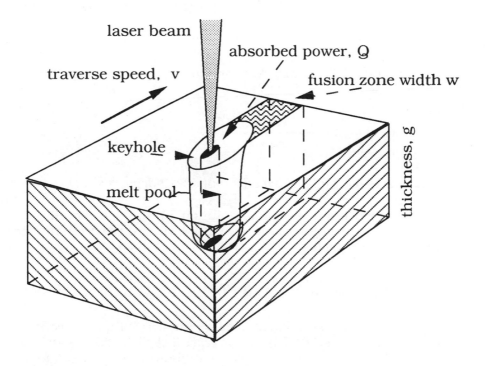

Fig. 5.8. The approximate geometry of the heat flow pattern around a keyhole weld.

and $-\dfrac{\partial T}{\partial R}\, 2\pi\, R\, k\, g \rightarrow Q$ as $R \rightarrow 0$

where R = distance from the heat source = $\sqrt{x^2 + y^2}$ m

T	=	Temperature at point x,y	K
T_o	=	Original plate temperature	K
Q	=	Heat input/unit time = $P(1-r_f)$	W
v	=	Welding speed	m/s
g	=	Plate thickness	m
k	=	Thermal conductivity	W/mK
α	=	Thermal diffusivity	m²/s
K_o	=	Bessel function 2nd kind and zero order	

(Bessel functions are sometimes known as cylinder functions and occur in diffusion problems with cylindrical symmetry and many other occasions. Bessel functions of the first kind known as a "Bessel function" are

$$J_p(z) = \sum_{n=0}^{\infty} [(-1)^n (\tfrac{z}{2})^{p+2n}] \, / \, [n!\,\Gamma\,(p+n+1)]$$

where $\Gamma(p+n+1) = (p+n)!$
Bessel functions of the second kind are known as a Nuemann functions:

$$Y_p(z) = \frac{\cos(p\pi)\,J_p(z) - J_p(z)}{\sin p\pi}$$

Such functions have tables available but are difficult to solve).

From equation 5.11 an asymptotic solution for the rate of cooling can be derived for the centreline location: $x > 0$, $y = 0$ and the Peclet number $vx/2\alpha \gg 1$ (i.e. at high speeds) as (10):

$$\frac{\partial T}{\partial t} = 2\pi\, k\, \rho\, C \left[\frac{vg}{Q}\right]^2 (T-T_0)^3 \qquad (5.15.)$$

It can be seen that the relationships so far derived are between the dimensionless groups: $[vR/2\alpha]$ and $[Q/\{2\pi(T-T_o)kg\}]$. Swifthook and Gick (11) observed that there were analytic solutions for the Bessel function at high or low speeds and hence made the plot shown in Fig. 5.9. The limit for high speed has already been mentioned in Chapter 4, Section 4.4.1. The limit for low speed is not so straightforward to understand.

This graph is useful in predicting required powers and speeds for welding.

For example: a laser beam is required to weld 10mm thick 304 stainless steel at 10mm/s. The usual weld width for the laser being considered is 1.5mm. What laser power is required assuming a 90% absorption in this keyhole welding process?

The thermal properties of 304 stainless steel are:

Thermal diffusivity, α	0.49×10^{-5}	m^2/s
Thermal conductivity, k	100	W/m/K
Melting point, T_m	1527	°C

Solution:
$Y = vw/\alpha = (0.01 \times 0.0015)/0.49 \times 10^{-5} = 3.06$
From Fig. 5.9 or from the equation $Y = 0.483 \, X$
$X = 7.0 = Q/gkT = Q/\, 0.01 \times 100 \times 1527$

$\therefore \; Q = 10.6kW$
with 90% transfer efficiency required power is $10.6/0.9 = \underline{11.8kW}$

This is a large power, but then it is capable of welding thick stainless

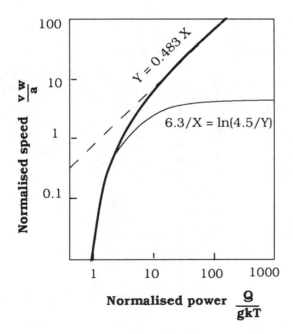

Fig. 5.9. Normalised speed vs normalised power. A solution of the moving line source problem (11).

steel at a high speed. If we only had a 2kW laser then the calculation would show what speed we could expect. That is an excercise for the reader! One can see that such sums assume that the penetration is possible and that the beam can be focussed to a fine spot - a line. Also that any calculation of the temperature near or within the experimental beam diameter will be badly wrong. These limitations can be overcome to some extent by using a cylindrical heat source as in Bunting and Cornfield's solution (12); or by using finite element or finite difference numerical methods.

5.7. Analytic Moving Point-Line Source Solution

Since temperatures can be added it is relatively simple to add the point and line source solutions and so simulate the Fresnel absorption by the line source and the plasma absorption by the point source located at any specified point along the line. Such a solution has been derived by Steen et al.(13).

Fig. 5.10. shows the fit of an experimental fusion zone to the calculated zone allowing one to speculate on the proportion of power in the plasma or Fresnel absorption.

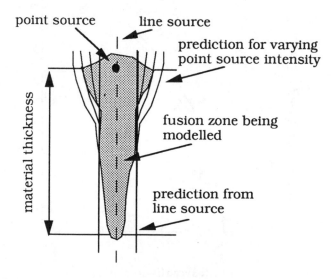

Fig. 5.10. Illustration of the solution achieved with the point line source model.

5.8. Finite Difference Models (14,15,16)

The finite difference or finite element models all solve the basic equations from Fourier's second law for all internal points and then have special equations for the boundaries. A finite difference model starts by dividing up the space to be considered into small boxes, for example, the model of Mazumder et al (14) uses an exponential grid for the first few calculations to get an estimate on how far the heat spreads, and therefore the volume that has to be considered. Since a regular matrix gives greater computing stability, once the size of the zone of interest is established the model changes to a regular matrix. For each of these boxes a heat balance is made. For example the heat balance on the element shown in Fig. 5.11 which is either heating up or cooling down, is given by Fourier's 2nd law as:

$$\alpha \frac{\partial^2 T}{\partial z^2} = \frac{\partial T}{\partial t} - u \frac{\partial T}{\partial z}$$

This can be expressed in finite difference form as:

$$\alpha \left[\frac{\dfrac{T_w - T}{\Delta z} - \dfrac{T - T_E}{\Delta z}}{\Delta z} \right] \approx \frac{T' - T}{\Delta t} - u \left[\frac{T_W - T_E}{2\Delta z} \right]$$

$$\therefore \ T' = \left\{ u \left[\frac{T_W + T_E}{2\Delta z} \right] + \alpha \left[\frac{T_W + T_E - 2T}{\Delta z^2} \right] \right\} \Delta t + T \qquad (5.16.)$$

convection and conduction

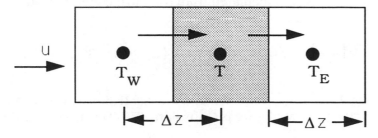

Fig. 5.11. Notation for equation 5.16.

The 3D version of this equation is solved for all internal points in order to find the temperature after one time interval, Δt. This is done all over the whole matrix. An instability may arise in that the two gradient terms are centred on time $t = t$, while the variation of temperature with time term is centred on $t = t + \Delta t/2$. The Crank Nicholson method (17) is sometimes used to overcome this problem which could otherwise be a source of computing inefficiency.

If the problem is one of a quasi-steady state nature, such as the weld pool beneath a laser beam in a moving substrate, then the time term may be dropped and the solution is found through a process known as "relaxation". In this form of calculation, used by Mazumder et al (14), the heat terms should add to zero. If they do not then the steady state temperature field is slowly relaxed, i.e. changed, by a calculated amount, until they do.

If there is melting then the latent heat is usually treated as an abnormal specific heat. Henry et al (15) arranged for stages in melting to be considered by letting the specific heat C_p be a step function of the form:

$$C_p(T) = C_{po} + \frac{\Delta H}{\Delta T_m} \qquad \text{from } T_m \leq T \leq T_m + \Delta T_{m'}$$

$$\text{and} \quad C_p(T) = C_{po} + \frac{\Delta H_v}{\Delta T_v} \qquad \text{from } T_v \leq T \leq T_v + \Delta T_{v'} \qquad (5.17.)$$

where C_{po} is the normal constant specific heat; ΔH_v, ΔH_m are the latent heats of boiling and fusion respectively and ΔT_v, ΔT_m are the temperature bands over which the transition occurs.
The boundary points have a similar heat balance to the internal points, but there has to be some modification to the gradients extending outside the zone, where the temperatures are not calculated.

The surface points are calculated from the surface temperature gradient:

$$k\left(\frac{\partial T}{\partial x}\right)_{x,y,1} = P_{x,y}(1 - r_f) - (h_c + h_r)(T_{surf} - T_a) \quad (5.18.)$$

The value of P_{xy} depends upon which power distribution is chosen. For example if the mode structure is Gaussian, TEM00, then P_{xy} is defined as:

$$P_{xy} = \frac{P_{tot}}{r_b^2 \pi} \exp\left(\frac{-2r^2}{r_b^2}\right) \tag{5.19.}$$

where $r^2 = x^2 + y^2$.

Boiling of the surface can be allowed for by assuming that if the boiling point is reached at a certain matrix point then that point will disappear allowing the power to fall on the matrix point beneath it, with some plasma absorption loss accounted for by Beer Lambert's absorption law (16):

$$P_{xy} = P_o e^{-\beta \Delta z} \tag{5.20.}$$

where β is an absorption coefficient with the units of m^{-1}

The side points are considered to be sufficiently far away that $\partial T/\partial x$ and $\partial T/\partial y \sim 0$: the base points for thick substrates are assumed to have $\partial T/\partial z \approx 0$; but the base points for thinner substrates are calculated from the estimated z gradient:

$$\left(\frac{\partial T}{\partial z}\right)_{x,y,lz} = -(h_c + h_r)(T_{surf} - T_a)/k \tag{5.21.}$$

Hopefully this quick summary has shown the reader, not already familiar with these techniques, how very versatile numerical models can be and that almost any physical phenomenon can be added and considered. This is clearly the strength of this technique as a tool for questioning process mechanisms. However, since the average model requires 0.5M bytes of memory and several minutes on a fast computer, it is not a model useful for every day prediction or control work. Some examples of the sort of solutions one can obtain from this form of analysis are shown in Figs. 5.12, 5.13, 5.14 and Fig. 4.31.

5.9. Semi-Quantitative Models

There are a number of these but the one by Klemens (20) should be noted. He was analysing the keyhole by assuming only radial heat flow, a heat balance indicates that:

$$P(r) = -k\left(\partial T/\partial r\right) 2\pi r \tag{5.22.}$$

If the power is only absorbed by the keyhole plasma of radius, r_c, then we know that for a "top hat" mode (uniform step function):

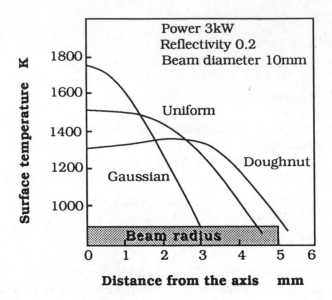

Fig. 5.12. Finite difference solution for the effect of mode structure on the surface temperature distribution (18).

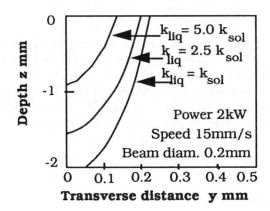

Fig. 5. 13. A finite difference solution illustrating the effect on the melt pool size of the effective thermal conductivity of the melt pool. The effective conductivity is made up of the material conductivity and a convective element or eddy conductivity. This calculation shows how significant the pool stirring action could be on the shape and size of the melt pool (19).

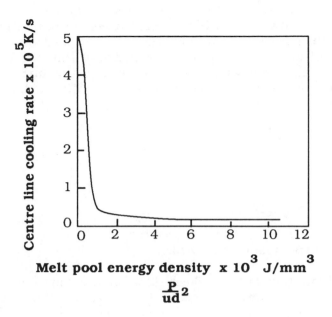

Fig. 5.14 A finite difference solution for the centreline cooling rate versus the melt pool energy density. The cooling rate will determine the structure of the solidification process (see chapter 6)

$$P(r) = W\pi r_c^2 = P \qquad\qquad (5.23.)$$

where W = Power/unit area/unit depth in $W/m^2/m$
Integrating (5.22.) and using (5.23.) as a boundary condition gives:

$$(P/\,2\pi k)\,[\,\ln(r/r_c)] = T_c - T \qquad\qquad (5.24.)$$

where T_c is the plasma temperature in K
At the edge of the keyhole $r = r_o$ and $T = T_v$ the boiling point, K:

$$\therefore\ r/r_c = \exp[(\,2\pi k\,(T_c - T_v)\,/\,P\,] \qquad\qquad (5.25.)$$

Thus we have some concept of the structure and temperatures within the keyhole. There are other gems in Klemen's paper which show an estimate of the size of the keyhole and that there is likely to be a neck at the top of it.

Other semi-quantitative arguments are presented in Steen and Courtney (21) showing that the transformation hardening depth is likely to be proportional to the group $P/\sqrt{(VD)}$. The reasoning is as follows.

The depth of penetration of a given isotherm will be governed by the laws of heat conduction which as we have seen involves the Fourier number. Thus the depth of this isotherm will be given from a particular value of the group $[\Delta x^2/\alpha t]$. The value of the isotherm at that depth will be governed by the heat intensity, which is P/D. Now the time of heating from a moving disc source will be given by some value such as (D/V). Assuming the relationships are linear we have:

$$\Delta x = A\left(\frac{P}{D}\right) . \left(\sqrt{\frac{\alpha D}{V}}\right)$$

and therefore

$$\Delta x = f\left(\frac{P}{\sqrt{DV}}\right) \qquad\qquad (5.26.)$$

which fits the experiments quite well.

A further illustration is from Steen (22) on the movement of the solidification front beneath a laser clad. It shows the likelihood that the interface region can be made to resolidify almost instantly after the formation of a fusion bond, if the operating conditions are correct.

5.10. Flow Models

Some brave souls such as Mazumder (23) have solved the Navier Stokes equations for the Marangoni driven flow within the melt pool. The melt pool size is calculated from the above finite difference method. Such solutions produce some beautiful graphs and illustrate that the melt pool is rotating vigorously. However they fail to account for internal turbulence required to explain the mixing action and so far have difficulty with describing the formation of waves. It is, however, early days.

5.11. Conclusions

The activity of modelling has given great confidence to our understanding of the process mechanics and hence made laser processing more acceptable as an established and reliable processing method.

5.12. List of Symbols

A	Area or constant	m^2
C_p	Specific heat at constant pressure	J/kgK
D	Beam diameter	m
F_o	Surface absorbed power	W/m^2
g	Thickness	m
H	Heat generated/unit volume	W/m^3
k	Thermal conductivity	W/mK
P	Beam power	W
q	Heat flow	W
Q	Absorbed power	W
R	Radial distance	m
r_f	Reflectivity	
r_b	Beam radius on workpiece	m
T	Temperature	K
t	Time	s
U	Velocity	m/s
u,v	Velocity, traverse speed	m/s
W	Power/unit area/unit depth	W/m^3
x	Distance in x direction	m
y	" y "	m
z	" z "	m
α	Thermal diffusivity	m^2/s
β	Beer Lambert coefficient	m^{-1}
ρ	Density	kg/m^3

subscripts

b	Beam condition
c	Critical condition
o	Start condition
p	Required value
tot	Total
v	Vaporisation, boiling condition
x,y,z	Vector direction or location
1	Finish condition
surf	Surface condition

References

1. Perry's Chemical Engineers Handbook, publ McGraw Hill 5th ed 1973 2-114.
2. Abramowitz.M, Stegun.I.A. eds 1964 Handbook of Mathematical Functions, National Bureau of Standards, Applied Mathematics Series No.55 (US GPO Washington DC).
3. Brienan.E.M. Kear.B.H. "Rapid Solidification Laser Processing at High Power Density" Chapter 5. Laser Material Processing ed M.Bass publ North Holland Publishing Co,Amsterdam, pp235-295 1983.
4. Carslaw.H.S., Jaeger.J.C. "Conduction of Heat in Solids" 2nd Edition Oxford University Press, 1959.
5. Rosenthal.D. Trans ASME 849-866 1946.
6. Ashby M.F., Shercliffe.H.R. "Master plots for predicting the case depth in laser surface treatments" CUED/C-MAT/TR134.
7. Ashby M.F., Easterling K.E. "The transformation hardening of steel surfaces by laser beams" Acta Met 32, 1935-1948 (1984).
8. Davis.M., Kapadia.P., Dowden.J., Steen.W.M., Courtney.C.H.G. "Heat hardening of metal surfaces with a scanning laser beam", J.Phys. D: Appl. Phys 19 (1986) 1981-1997.
9. Bradley.J.R. "A simplified correlation between laser processing parameters and hardened depth in steels"J.Phys D: Appl. Phys. 21 (1988) 834-837.
10. Dowden.J., Gratzke.U., Essex University, private communication Jan 1991.
11. Swifthook.D.T., Gick.E.E.F. Welding Res Suppl 492s-498s Nov 1973.
12. Bunting. K.A., Cornfield. G. 1975. Trans ASME J.Heat.Trans. (Feb) p.116.
13. Steen.W.M., Dowden.J, Davis.M., Kapadia.P, "A point line source model of laser keyhole welding" J.Phys D appl. 21 1988 pp1255-1260.
14. Mazumder.J., Steen.W.M. "Heat Transfer Model for CW laser Processing" J.App.Phys Feb 1980 51 (2) pp941-946.
15. Henry.P., Chande.T., Lipscombe.K., Mazumder.J., Steen.W.M. "Modelling laser heating effects" paper 4B-2 proc int Conf on Laser and Electro optics ICALEO '82 Boston Sept 1982 publ LIA Tulsa USA.
16. Eboo.M. PhD thesis, London University 1979.
17. Crank.J., Nicholson.P. Proc Camb. Phil. Soc. 43 1947 50-67.
18. Sharp.M. Ph.D. Thesis, University London 1986.
19. Steen.W.M., Mazumder.J. "Mathematical modelling of laser/ material interactions" Report AFOSR-82-0076 GRA vol 84 NO.17 Aug 17th 1984.

20. Klemens.P.G. 1976 J.App.Phys (May) 47 No5 2165.
21. Steen.W.M. & Courtney. G.H.C. 1979 Metals Technology 6
    No 12, 456.
22. Steen. W.M., Powell. J., Proc.Eur. Scientific. Laser Workshop
    '89. publ. Sprechsaal publ.Co., Coburg FRG 1989, pp. 143-160.
23. Chan.C., Mazumder.J., Chen.M.M. "A two dimensional
    Transient Model for Convection in a laser Melted Pool"
    Met Trans A vol 15A Dec 1984 pp2175-2184.

"A Fast Fourier Transformation!"

Chapter 6

Laser Surface Treatment

"Beauty is only skin deep, but it is only the skin you see"
15th century proverb amended by A.Price 44 Vintage xix 1978.

6.1. Introduction

The laser has some unique properties for surface heating. The electromagnetic radiation of a laser beam is absorbed within the first few
atomic layers for opaque materials, such as metals, and there are no
associated hot gas jets, eddy currents or even radiation spillage outside
the optically defined beam area. In fact the applied energy can be placed
precisely on the surface only where it is needed. Thus it is a true surface
heater and a unique tool for surface engineering. The range of possible
processes with the laser is illustrated in Fig. 6.1. Common advantages
of laser surfacing compared to alternative processes are:

* Chemical cleanliness.
* Controlled thermal penetration and therefore distortion.
* Controlled thermal profile and therefore shape and location of heat
  affected region.
* Less after machining, if any, is required.
* Remote non contact processing is usually possible.
* Relatively easy to automate.

Surface treatment is a subject of considerable interest at present
because it seems to offer the chance to save strategic materials or to
allow improved components with idealised surfaces and bulk properties. These ambitions are real and possible but economically it is likely
that this current enthusiasm will fade into a realisation that only

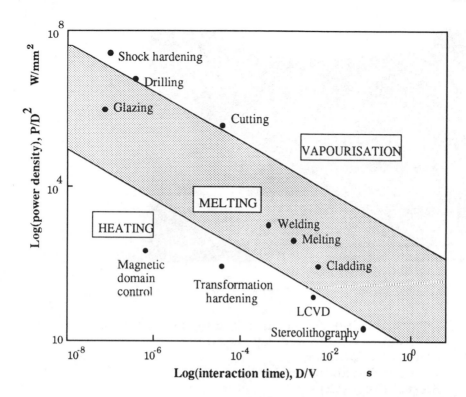

Fig. 6.1. Range of laser processes mapped against power density per unit time.

particular parts of surfaces are vulnerable to corrosion or wear and that there is no need to cover great areas. Where great areas are required to be covered, for example for appearance, paint will probably be most cost effective, or for large area coverage by metals then electroplating is likely to be the winner; but for discrete areas the laser has few competitors and can give a wide variety of treatments as discussed here.

Currently the uses of lasers in surface treatment include:

* Surface heating for **transformation hardening** or annealing.
* Surface **melting** for homogenisation, microstructure refinement generation of rapid solidification structures and surface sealing.
* Surface **alloying** for improvement of corrosion, wear or cosmetic properties.
* Surface **cladding** for similar reasons as well as changing thermal properties such as melting point or thermal conductivity.
* Surface **texturing** for improved paint appearance.
* **Plating** by Laser Chemical Vapour Deposition (LCVD), Laser Physical Vapour Deposition (LPVD),

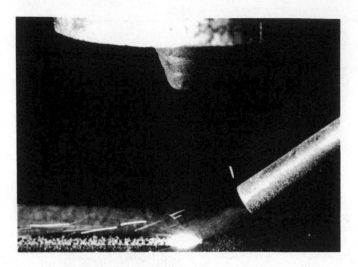

Fig. 6.2. A blown powder laser clad being formed.

Enhanced Plating for localised plating by electrolysis or
cementation or improved deposition rates.
*   Non-contact **bending**.
*   Magnetic **domain control**.
*   **Stereolithography**.
*   **Paint stripping**.

It can be seen that these processes range from the low power density
processes of transformation hardening, bending and laser chemical
vapour deposition (LCVD) which rely on surface heating without surface
melting, to processes involving surface melting requiring higher power
densities to overcome latent heat effects and larger conduction heat
losses; these melting processes include simple surface melting to
achieve greater homogenisation, or very rapid self quench processes as
in laser glazing for the formation of metallic glasses, which is possible
in certain alloys. Melting processes also include those where a material
is added either with a view to mixing into the melt pool as in surface
alloying and particle injection or with a view to fusing on a thin surface
melt, as in cladding (Fig. 6.2). If very short pulses of great power intensity
strike a surface they are able to send mechanical shock waves, originat-
ing from sudden thermal stress, through the material resulting in
surface hardening similar to shot peening but with a greater depth of
treatment. The processes listed above are now discussed in turn.

1. Heating Processes

	Advantages	Disadvantages
Laser	Minimum part distortion Selective hardening No quenchant required Thin case capability Case depth controllable Eliminates post processing Improves fatigue life	High equipment cost Coverage area restricted Absorbent coatings necessary Multiple passes give local tempering
Induction	Fast process rates Deep case obtainable Lower capital cost than laser Coverage area	Downtime for coil change Quenchant required Part distortion Coil placement critical Large thermal penetration EM forces may spoil surface Fabrication of complex coils for specific processes
Flame	Cheap , flexible and mobile process	Poor reproducibility Lacks rapid quench Component distortion likely Environmental problems
Arc (TIG)	Relatively cheap and flexible process	Section thickness limited Large thermal penetration Stirring takes place Poor control to avoid melting
Electron beam	Minimal distortion, selective hardening and no quenchant required	High equipment cost Requires vacuum Low production rate High processing costs

2. Thermochemical Diffusion Processes

Carburising — Gas, Liquid, Pack — Conventional < 950 $^{\circ}$C, High temperature > 950 $^{\circ}$C, Low pressure ("vacuum"), N_2 based carrier gas, Fluidised bed

Nitriding — Gas — Conventional, Ion nitriding

Carbonitriding (austenitic) — Gas, Liquid

Fig. 6.3. Chart of Competing Processes for Heat Treatment.

6.2. Laser Heat Treatment (1)

The initial goal of laser heat treatment was selective surface hardening for wear reduction; it is now also used to change metallurgical and mechanical properties. There are many competing processes in the large subject of surface heat treatment, Fig. 6.3. The laser usually competes successfully due to lack of distortion and high productivity. Practical uses of laser heat treatment include:

*   hardness increase.
*   strength increase.
*   reduced friction.
*   wear reduction (2).
*   increase in fatigue life.
*   surface carbide creation.
*   creation of unique geometrical wear patterns.
*   tempering is also possible.

Laser heat treatment is used on steels with sufficient carbon content to allow hardening and cast irons with a pearlite structure. The process arrangement is illustrated in Fig. 6.4. An absorbing coating is usually applied to the metal surface to avoid unnecessary power loss by reflection. Some typical coatings with the average reflectivity values are listed in Table 6.1. The absorption coefficient can also be increased by allowing a polarised beam with the electric vector in the plane of incidence, to be reflected at the Brewster angle (approximately 8° for metals) (3). This leads to a unique process for transformation hardening inside small holes - such as valve guides. The variation of reflectivity with angle of incidence has been noted in Chapter 2 Section 2.3.4. As the beam moves over an area of the metal surface, the temperature starts to rise and thermal energy is conducted into the metal component. Temperatures must rise to values that are more than the critical transformation temperature (Ac1) but less than the melt temperature. After the beam has passed cooling occurs by quenching from the bulk of the material which has hardly been heated by this fast surface heating process. The thermal and structural histories are illustrated in Figs. 6.5, 6.6a and 6.6b. The process is discussed later in Section 6.2.3 but in essence for transformable alloys there is a phase change on heating which starts at the Ac1 temperature of Fig.6.6a and is complete at the Ac3 temperature. This new structure is unable to transform back again on rapid cooling due to diffusion which occurs while at the higher temperatures. The species diffusing is usually carbon. The result is a structure under some form of stress and hence unable to allow dislocations to flow. Such a structure has the property of being hard.

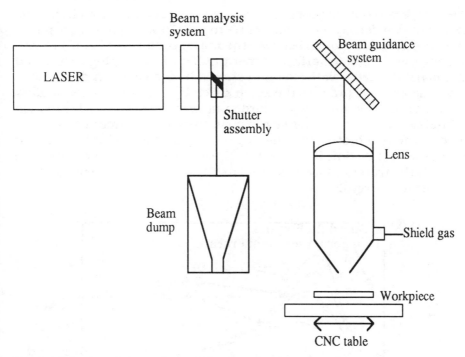

Fig. 6.4. Experimental arrangement for laser heat treatment.

Table 6.1	Typical values of the reflectivity of various surfaces to 10.6μm radiation at normal angles of incidence.		
Surface Type	Reflectivity %		
	Direct	Diffuse	Total
Sandpaper roughened (1μm)	90.0	2.7	92.7
Sandblasted (19μm)	17.3	14.5	31.8
Sandblasted (50μm)	1.8	20	21.8
Oxidised	1.4	9.1	10.5
Graphite	19.1	3.6	22.7
Molybdenum sulphide	5.5	4.5	10.0
Dispersion paint	0.9	0.9	1.8
Plaka paint	0.9	1.8	2.7

The laser beam is defocussed or oscillated to cover an area such that the average power density has a value of 10^3 to 10^4 W/mm². Using these power densities a relative motion between the workpiece and the beam of 5 and 50 mm/s will result in surface hardening. If surface melting occurs and this is not desired, relative motion should be increased. A decrease in power density will produce the same effect. If no hardening, or shallow hardening occurs, but deeper hardening was desired, relative motion should be decreased, an increase in power density will produce the same effect. The depth of hardening depends upon thermal diffusion and hence the heating time (D/V); where D is the spot size on the workpiece and V is the traverse speed, as well as the temperature, dependent on the specific energy (P/DV).

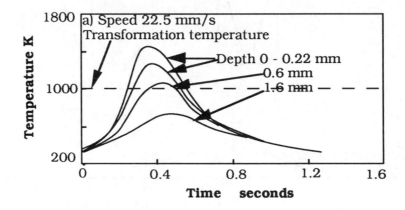

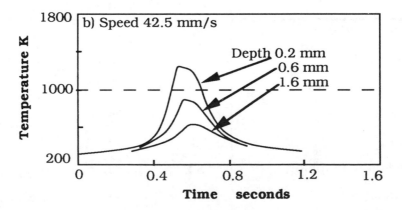

Fig. 6.5. Theoretically predicted thermal cycles during laser heating of En8 steel (power = 2kW, beam radius = 3mm and reflectivity = 0.4) (4).

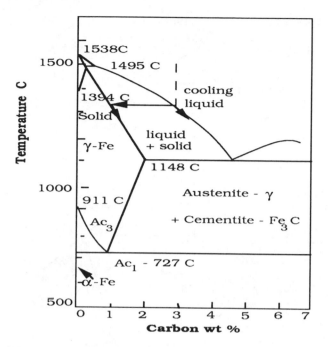

Fig.6.6. (a) The iron cementite (Fe –Fe$_3$C) system – a summary of stable and metastable equilibria to 7wt% C (5).

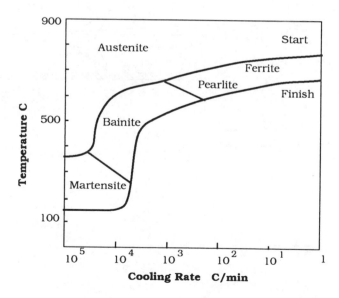

Fig. 6.6. (b) Continuous cooling transformation diagram for a 0.38 wt% C steel (6). Analysis: C 0.38; Si 0.20; Mn 0.70; P 0.02; S 0.02.

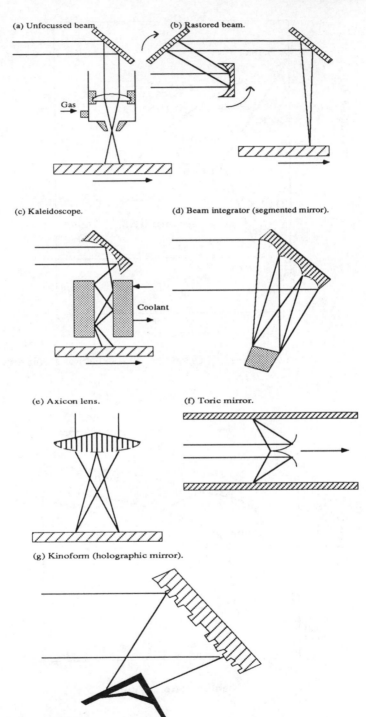

Fig.6.7. Methods of beam spreading.

6.2.1. Heat Flow

The ideal power distribution is one which gives a uniform temperature over the area to be treated. This requires a dimpled power distribution since the heating effect is dependent on the edge cooling and surface heating, i.e. P/D and not P/D^2 where P is the incident absorbed power. Methods of spreading the beam to simulate this are illustrated in Fig. 6.7. They include:

i).   Defocussed high power multimode beams (top hat mode)
ii).   One or two axis scanning beams (dithered zig zag mode)
iii).   Kaleidoscopes
iv).   Segmented mirrors
v).   Special optics (axicon lenses, toric mirrors and kinoforms)

Most of these are used in laser heat treatment. All of these generate a reasonably uniform distribution of power over the central region of the beam path. The temperature distribution with depth during the temporal duration of the irradiation can be represented by equations derived from simple, but idealised models of one-dimensional (1D) heat transfer. A simple test to determine if this representation can be used is to examine the cross-section of a heat treated sample as in Fig. 6.8. If the bottom of the hardened zone is flat and parallel to the surface under the central part of the cross-section, then the one-dimensional analysis will predict the temperatures in the heated material with reasonable accuracy (as discussed in Chapter 5). The edges of the cross-section are regions where the problem is two-dimensional (2D) and the one dimensional heat flow model will not accurately predict the induced temperatures. Whether the edge or central model is dominant is determined by the processing speed and beam diameter expressed as the Peclet number $(Dv\rho C/k)$ (7). Transformation hardening with no surface melting is the simplest process to model mathematically (1);   there are no unknown convection or latent heat terms since there is no melt pool and surface heat losses follow the normal rules of convection and radiation (8,9). An empirical relationship between $P/(DV)^{1/2}$ and the depth of hardness was found by Courtney (4). The theoretical fit for this parameter, as calculated by Sharp (7) using a finite difference model, is shown in Fig. 6.9.   The spread due to the Peclet number effect is shown to be slight. For En8 the Courtney fit was found to be:

$$d = -0.10975 + 3.02\frac{P}{\sqrt{DV}} \qquad\qquad (6.1)$$

Although the 1D distribution is useful for approximate predictions, if

Fig. 6.8. Microstructure of laser transformation hardened En24 steel. Power 1.6kW, traverse speed, 15mm/s, beam diameter, 6mm.
Composition (wt %): C 0.36-0.44; Si 0.1-0.35; Mn 0.45-0.7; P 0.035 max; S 0.04 max; Cr 1.0-1.4; Mo 0.2-0.35; Ni 1.3-1.7.   **×50**

more exact thermal distributions are required then calculation must be made via numerical techniques such as finite difference models (10). This would be the case if allowance is to be made for variations in beam energy distribution, edge effects, finite part size and particular geometries. It is also true if the hardened width is an essential part of the answer sought.

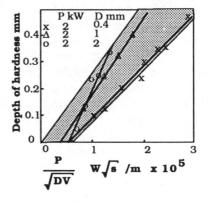

Fig. 6.9. Theoretical plot of P/√(DV) for the depth of the 1960K isotherm for various P,V and D (from 7).

A greatly improved model has been suggested by Ashby and Shercliffe (11) who solved the problem of a moving finite line source placed at a specified distance above the surface
- to simulate a beam diameter. They produced the master plot illustrated in Fig. 5.7. This plot has been shown to fit the results of transformation hardening quite well.

6.2.2. Mass Flow by Diffusion

In transformation hardening of steels, the parent structure consists of a non-homogeneous distribution of carbon, e.g. pearlite and ferrite, which upon heating above the phase transformation temperature, Ac1 temperature (723°C for many steels), starts to diffuse to achieve homogeneity within the austenite phase. The rate of diffusion is described by similar equations to that for heat flow, but is usually much slower, i.e.:

$$\frac{\delta c}{\delta t} = D_{AB} \left[\frac{\delta^2 c}{\delta x^2} + \frac{\delta^2 c}{\delta y^2} + \frac{\delta^2 c}{\delta z^2} \right] \tag{6.2}$$

The diffusivity of carbon in austenite is approximately

$$D = 1 \times 10^{-5} \, e^{-9.0/T} \quad m^2/s$$

and in ferrite

$$D = 6 \times 10^{-5} \, e^{-5.3/T} \quad m^2/s.$$

When austenitisation has occurred the carbon moves by diffusion down concentration gradients. The time for diffusion within the austenitic lattice varies with position within the laser treated zone (7), Fig. 6.5. In laser transformation hardened zones there is always a region around the edges, if not throughout, where the carbon has not fully diffused and the resulting structure is a non-homogeneous martensite; it is not yet known whether this non-homogeneous martensite is preferable to homogeneous martensite. It would be expected that the higher carbon levels in certain regions would lead to higher hardness levels and therefore better overall wear resistance.

6.2.3. Mechanism of Transformation Process

6.2.3.1. *Steels:* On rapid heating, pearlite colonies first transform to austenite. Then carbon diffuses outwards from these transformed zones into the surrounding ferrite increasing the volume of high carbon austenite. On rapid cooling these regions of austenite which have more than a certain amount of carbon (e.g. 0.05 %) will quench to martensite if the cooling rate is sufficiently fast although retained austenite may be found if the carbon content is above a certain value (>1.0 %). The required rate of cooling is indicated by constant cooling curves, Fig. 6.6b. In laser transformation hardening the cooling rate is usually in

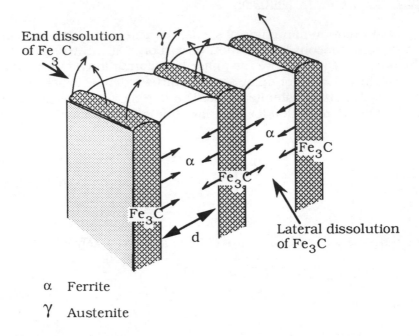

α   Ferrite

γ   Austenite

Fig. 6.10. Routes for carbon dissolution during homogenisation (12).

excess of 1000°C/s which means that most steels will self quench to martensite not bainite or pearlite.

The transformation of the pearlite is thought to proceed by diffusion from the cementite plates into the ferrite plates, possibly starting from one end of a pearlite colony, Fig. 6.10. This time dependent process does not take long but is sufficient to necessitate some superheat above the austenitising temperature, Ac1, to allow it to proceed to any extent during laser treatment. The superheat, and therefore the extent of the diffusion process, is thus slightly affected by the prior size of the pearlite colonies. These colonies, on transformation, become austenite having 0.8 % carbon. Carbon diffuses down the concentration gradient into the ferrite regions where there is virtually no carbon. The ferrite regions may also have transformed to the fcc (face centred cubic) structure of austenite. The extent of homogeneity of the resultant martensite will depend upon the size of the prior ferrite regions and the processing conditions- in particular the interaction time (beam diameter/traverse speed). The hardness depends upon the carbon content, Fig. 6.11.a,b.

The metallurgical changes which occur in laser treated steels are similar

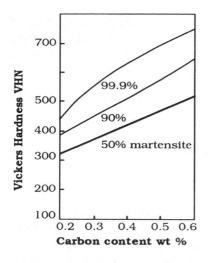

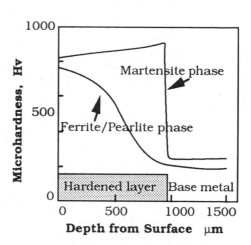

Fig. 6.11a. Average relationship between carbon content and hardness for
steels containing different amounts of
untempered martensite (13).

Fig. 6.11b. Micro hardness distribution in
a nonhomogeneous hardened specimen (14).

to those for furnace treated steels. However, the more rapid heating and
quenching of the laser process does result in variations in the type of
martensite, particularly its fineness, amount of retained austenite and
carbide precipitation as well as the homogeneity of the hardened zone.
The transformed zone is also more highly restrained resulting in higher
compressive stresses opposing the approximately 4% volume increase
associated with martensitic phase changes.

6.2.3.2. Cast Irons: Ferritic grey cast iron consists of ferrite and graphite
regions. As such it is difficult to harden by the laser because the
diffusion time is too short. Typically the diffusion distance from the
graphite is 0.1 mm for a 5 mm beam travelling at 20 mm/s. Thus all that
is formed is a hard crust around the graphite flakes or nodules. These
can still give impressive wear properties though no change in the overall
hardness value would be observed.

Pearlitic cast iron, formed by moderately fast cooling, consists of pearlite
and graphite. In this case laser transformation hardening is successful
in achieving very high hardness levels, as for 0.8 % C irons or higher.
With cast irons there is a fairly narrow window between transforming
and melting. The irons are important for their ease of casting, hence they
have low melting points, while their Ac1 temperature is approximately
constant as for all Fe/C alloys (Fig. 6.6.a).

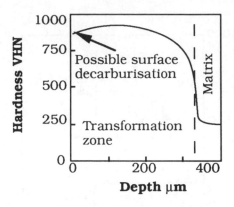

Fig. 6.12. Variation of micro hardness with depth for a 20CrMo steel.

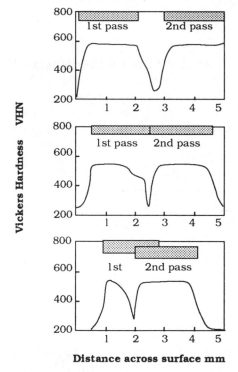

Distance across surface mm

Fig. 6.13. Plot of surface hardness versus width for variations in overlap between successive passes (from 12).

Laser transformation hardening of spheroidal graphite cast iron may result in preferential melting around the graphite nodules due to the lowering of the melting point as the carbon diffuses away from the graphite.

6.2.4. Properties of Transformed Steels

6.2.4.1. Hardness: This depends upon the carbon content (Fig. 6.11). It has been found that the hardness value may be slightly higher than that found for induction hardening. This difference is probably due to the shallower zone in the laser process allowing a faster quench and therefore greater restraint and hence higher residual compressive stress. A typical hardness profile is shown in Fig. 6.12 for a carburised 20 CrMo steel.

Overlapping successive tracks induces a thermal experience in the neighbouring tracks so that there is some back tempering. This is not necessarily undesirable since it allows space for oil and wear debris. The extent of the hardness variation is illustrated in Fig. 6.13. Patterned hardened surfaces have not received too much attention as wear surfaces mainly because prior to the laser they were difficult to make. The laser can make patterned surfaces easily and therefore opens a whole new study in tribology.

The tempering of steels by laser is used in the production of reinforcing wire for tyres. The wire is passed at speed through a shaped laser beam.

Table 6.2.	Experimental results of rotational wear resistance for transformation hardened SK5 (AISI WI) steel.	
Property	Method of transformation hardening	
	Laser	Induction
Hardness	HRC 64-67	HRC 60-63
Case depth	0.7-0.9 mm	2-3 mm
Load	101kg/mm2	101kg/mm2
Scuffing	no occurrence	slight
Wear loss	0.5	1.0

6.2.4.2. Fatigue: In steels and cast irons there is a residual compressive stress on transformation hardening due to the volume expansion on the formation of martensite (approximately 4% for 0.3 wt%C steel). This effect is particularly pronounced in the shallower hardened zones formed with the laser due to the greater restraint for such treatment. Fatigue cracks are generally initiated at the surface by tensile stresses; thus the fatigue load must be sufficient to overcome this residual compressive stress before a crack can propagate. Improved fatigue life compared to induction hardening has been reported with laser heat treatment (1,15).

6.2.4.3. Wear resistance: Wear resistance has also been found to improve with laser treatment compared to oil or water quench. On SK5 steel the pin on disc wear resistance of laser treated surfaces was found to be twice that of an induction hardened surface, Table 6.2.

6.3. Laser Surface Melting

For surface melting the experimental arrangement is similar to that for transformation hardening, shown in Fig. 6.4, except that in this case a focussed or near focussed beam is used. The surface to be melted is shrouded by an inert gas. The competing processes are listed in Fig. 6.14. The main characteristics are:
*   Moderate to rapid solidification rates producing fine near homogeneous structures.
*   Little thermal penetration, resulting in little distortion and the possibility of operating near thermally sensitive materials.
*   Surface finishes of around 25 μm are fairly easily obtained, signifying reduced work after processing.
*   Process flexibility, due to software control and possibilities in automation.

Process	Characteristics
Laser ————————	High capital cost, localised heating of sample, rapid solidification of melt zone to give fine recrystallised grain structure, and good homogeneity. Controllable surface roughness. Power density $10^2 - 10^4$ W/cm^2.
Flame ————————	Cheap capital cost, poor reproducibility, no fast quench available, environmental problems and sample distortion . A flexible and mobile process.
Plasma ———————	Medium capital cost, very low heat input to sample.
TIG———————————	Limited section thickness, electromagnetic stirring, weld bead may be rough, large thermal penetration and medium heat input.
Induction ———————	Cheaper than laser, large thermal penetration, electromagnetic force may spoil surface, fast processing rate, deep case possible and fast area coverage.

Fig. 6. I 4.Competing surface melting processes.

The main areas of variation in processing centre around controlling the reflectivity, shaping the beam and shrouding the melt pool. Reflectivity is difficult to control due to the melting process itself causing variations in the surface reflectivity. The initial reflectivity can be controlled in the same manner as for transformation hardening, by having an anti-reflection coating, but this is usually removed by the melting process; however, once the material becomes hot the reflectivity is reduced, due to the increased phonon concentration, as is indicated in the Bramsen equation (6).

$$\varepsilon_\lambda = 0.365 \left(\frac{\rho_r(T)}{\lambda} \right)^{1/2} - 0.667 \frac{\rho_r(T)}{\lambda} + 0.006 \left(\frac{\rho_r(T)}{\lambda} \right)^{3/2} \quad (6.3)$$

where: $\rho_r(T)$ = Electrical resistivity at a temperature T$^\circ$C.
         $\varepsilon_\lambda(T)$ = Emissivity at T$^\circ$C to radiation of wavelength, λ.
         λ = Wavelength of incident radiation, m.

The reflectivity varies with the angle of incidence (17) and surface films play a significant role. The small addition of oxygen to the shroud gas has a notable effect on reflectivity (18), see Fig. 4.26, Section 4.4.8.2. A surface plasma will initially help to couple the beam into the surface. If

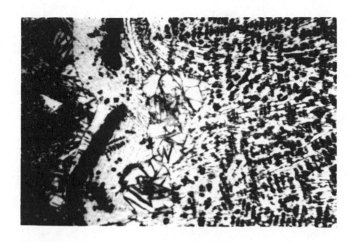

Fig. 6.15. Micrograph of the melt interface for laser surface melted flake graphite cast iron. X150

the plasma leaves the surface then it will block the beam. Optical feedback systems, such as a reflective dome around the interaction region (see Fig. 6.28), (19,20) can increase the laser coupling by around 40%. Optical methods vary according to the method uscd to produce the required spot size or beam shape which may be required to control the flow in the melt pool, as well as for the method used to protect the optics from sputter and fume.

There are three metallurgical areas of considerable interest; cast irons, tool steels and certain deep eutectics which can form metallic glasses at high quench rates. All are essentially non-homogeneous materials which can be homogenised by laser surface melting.

There are two reasons why laser surface melting is not widely used in industry:
i). If surface melting is required then surface alloying is almost the same process and offers the possibility of vastly improved hardness, wear or corrosion properties.
ii). The very high hardnesses achieved with cast irons and tool steels by laser surface melting are associated with some surface movement and hence may require some further surface finishing after treatment. This is not so easy to effect with the high hardnesses obtained.

The products of laser surface melting of some important engineering

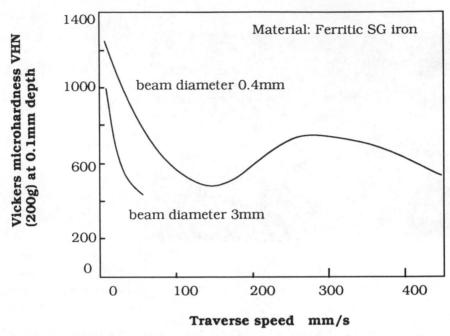

Fig. 6.16. Variation of microhardness with scanning speed for ferritic S.G. cast iron
(from 22).

materials are as follows:

Cast iron (21): This commonly used engineering material usually consists of an inhomogeneous structure of ferrite and graphite in various
forms (flakes, spheres etc). On surface melting with a laser the hardening
effects come from changes of graphite-to-cementite and austenite-to-
martensite (21-24). The precise value of the hardness depends on the
extent of the carbon dissolution from the graphite giving a variation of
hardness and structure with processing speed. The result is usually a
very hard surface on one of the cheaper metals and this can be achieved
by a simple, fast process. Fig. 6.15 shows the melt interface for laser
surface melted flake graphite cast iron. The structure varies from Fe_3C
dendrites in the ledeburitic fusion zone through high dissolved carbon
(around 1wt%) giving retained austenite with some martensite into full
martensite and partially dissolved graphite flakes. A form of trip
through the iron carbon phase diagram! Fig. 6.16 and 6.17 illustrate the
variation in hardness and structure with traverse speed. The high
hardness at slow speeds in Fig. 6.16 for SG iron is due to nearly all the
carbon dissolving giving a ledeburitic white iron structure. The second
peak at higher speeds, with only a small amount of carbon dissolution,

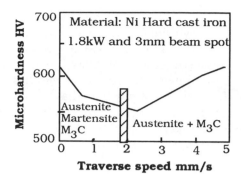

Fig. 6.17. Variation in microhardness with scanning speed for Ni-hard cast iron (21).

is due to a martensitic structure. The intermediate low hardness region is due to retained austenite. The improved wear properties are illustrated in Fig. 6.18 whereas the fatigue properties are usually worse due to residual tensile stress in non martensitic materials (12).

Stainless steel: Fine structures are produced in both martensitic or austenitic stainless steels as expected from the high values of the cooling rate, G.R. Without the phase expansion associated with the martensitic transformation, austenitic steels have a residual tensile stress while single tracks of martensitic steel are usually under compression, which becomes tensile when annealed by overlapping. The residual tension adversely affects the stress corrosion properties and the pitting potential (25). Lumsden et al (26) found that laser melting and rapid solidification had differing effects on the pitting behaviour of a series of ferritic steels of composition Fe-13Cr-xMo, where x varies from 0 to 5%. Unless the Mo concentration is higher than 3.5% laser melting had a deleterious effect or no effect on the pitting potential. The 5% alloy had a large increase

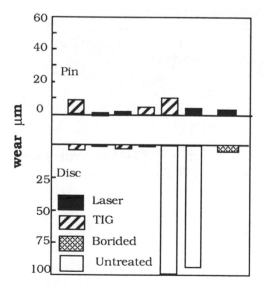

in the pitting potential compared to the untreated alloy. Improved corrosion resistance of sensitised stainless steels has been noted by many workers due to the finer structure reducing the tendency to intergranular corrosion (27).

Titanium: Titanium and its various alloys can take up a variety of crystal forms. In laser surface melting rapid quench structures are formed which have a highly dislocated fine structures (Fig. 6.19). The process must be carefully shrouded due to the activity of titanium with oxygen (29).

Fig. 6.18. Comparison of the wear of surface melted S G iron prepared by different processes (from 12).

Tool and special steels: These materials are usually hardened by a fairly long process of solution treatment to dissolve the carbides followed by a controlled quench to give a fine dispersion of carbides. These carbides do not temper as easily as martensite, hence these steels have a high hot hardness and are suitable for tools. In laser surface melting this dissolution is accomplished very swiftly producing a very hard, fine carbide dispersion with high hot hardness properties. The problem with the application of this process in production is that the laser melt track will have a surface waviness of around 10-25µm and the track is very hard to machine.

In all materials there is a tendency to cracking if the hardness is high. Usually this can be avoided if some preheat is applied. As a rule of thumb the required preheat is around 1°C/Vickers Hardness Number (VHN). This indicates a preheat of around 500°C for low carbon steel, 650°C for 0.7 wt%C steel and 700°C for tool steels.

6.3.1. Solidification Mechanisms

6.3.1.1. Style of Solidification (30):   Solidification will proceed as either a stable planar front or as an unstable front leading to dendrites or cells. The process which will occur depends on the occurrence of constitutional supercooling (Fig. 6.20. a,b,c,d). Constitutional supercooling is caused by the thermal gradient being less steep than the melting point gradient, which is the result of partition effects taking place at the solidification front giving rise to composition variation in this region.

Consider a mass balance on the solidification front, the gradient of the solute in the liquid at the solidification interface is:

$$\left[\frac{dC_L}{dx}\right]_{x=0} = -\frac{R}{D_L}C_L^* (1-k) \tag{6.4}$$

Constitutional supercooling is absent when the actual temperature gradient in the liquid at the interface, $G \geq (dT_L/dx)_{x=0}$; now the value of this gradient is:

$$\left[\frac{dT_L}{dx}\right]_{x=0} = m_L\left[\frac{dC_L}{dx}\right]_{x=0} \leq G \tag{6.5}$$

where: C_L = Liquidus composition,
          x = Distance from the interface,   m,
          T_L = Liquidus temperature,   °C,
          R = Rate of solidification,   m/s,

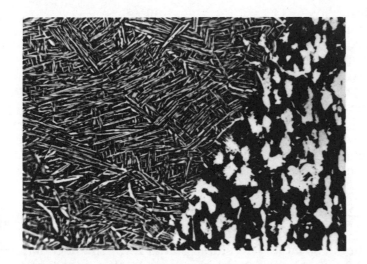

Fig. 6.19. Micrograph showing the fine basket weave structure produced in laser surface melting IMI550 (28) (P = 1.6kW; V = 200mm/s; D = 0.5mm). **X 100**

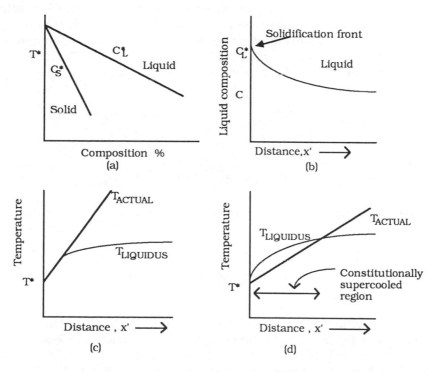

Fig. 6.20. Constitutional supercooling in alloy solidification (30). (a) phase diagram; (b) solute enriched layer in front of liquid-solid interface; (c) stable interface; (d) unstable interface.

D_L = Diffusivity, m^2/s,
C_L^* = Liquidus composition
          in equilibrium with solidus
          composition, C_s^*,
k   = Partition coefficient,
m_L = Slope of the liquidus, $[dT_L/dC_L]$,
G   = Thermal gradient,   °C/m.

Combining these two equations and letting $C_s^*=kC_L^*$ - the equilibrium condition - we obtain the general constitutional supercooling criterion: There is no constitutional supercooling if:

$$\frac{G}{R} \geq -\frac{m_L C_s^* (1-k)}{kD_L} \qquad (6.6)$$

The ratio (G/R) should be large for a stable planar front solidification mechanism. Fig. 6.21 illustrates this equation and further introduces the concept of "absolute stability" when the solidification rate, R, is so large that there is insufficient time for diffusion.

6.3.1.2. Scale of solidification structure: If the dendritic or cellular structure is sufficiently fine then it is possible to approximate the liquid between the cells as being like a small stirred tank whose composition will be determined by the rate of diffusion out of the cell depleting the concentration of the cell, in fact Fick's Second Law:

$$D_L \frac{\delta^2 C_L}{\delta y^2} = \frac{\delta C_L}{\delta t} \qquad (6.7)$$

Now:

$$dC_L/dt = (dC_L/dT)(dT/dx)(dx/dt) = -GR/m_L$$

Substituting for (dC_L/dt) and integrating across the cell width, λ, we obtain:

$$\left[\frac{\delta C_L}{\delta y}\right]_{y=0} = -\frac{GR\lambda}{m_L D_L} \qquad \text{and} \qquad \Delta C_{L_{max}} = -\frac{GR\lambda^2}{2m_L D_L} \qquad (6.8)$$

We observe that the parameter (GR) is related inversely to the square of the cell spacing, λ. (GR) is the cooling rate in °C/sec. In laser surface melting extremely high cooling rates can be achieved (~ 10^6 °C/sec) and therefore finer structures result, as illustrated in Fig. 6.22.

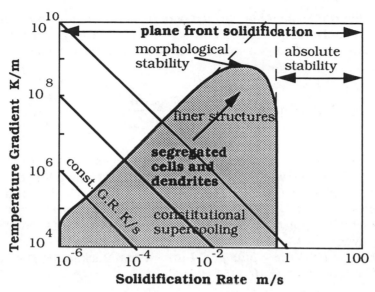

Fig. 6.21. Plot of temperature gradient versus solidification rate and solidification morphology.

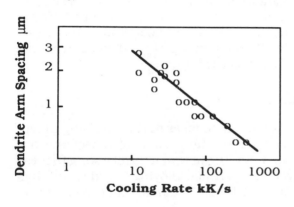

Fig. 6.22. Plot of dendrite arm spacing versus the logarithm of the cooling rate (GR).

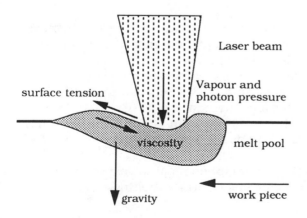

Fig. 6.23. forces on the melt pool.

6.3.1.3. Material flow within the melt pool: There are many forces acting on the melt pool as shown in Fig. 6.23.(31). One of the largest is that from the variation in surface tension, σ, due to the steep thermal gradients. Consider the following example:

$$\text{Surface shear force} = \frac{\delta\sigma}{\delta x} = \frac{\delta\sigma}{\delta T}\frac{\delta T}{\delta x} \qquad (6.9.)$$

For Ni the variation of surface tension with temperature, $d\sigma/dT = 0.38$ ergs/°C/cm² and for laser processing the thermal gradient is of the order of, $dT/dx \sim 2.5 \times 10^4$ °C/cm. Therefore shear force $= 0.38 \times 2.5 \times 10^4$ $= 10^4$ dynes/cm $= 10^3$ Pa $= 0.01$ atm. This is not an insignificant force but one which represents an acceleration of around 10G, on a small surface layer !
(F = ma; if F = 10^3 N/m² = ρta; if t = 1 μm - 1mm thick layer; ρ = 10,000 kg/m³, therefore a ~ $10^2 10^5$ m/s² ~ 10G).

Mazumder (10) has modelled this flow by solving the Navier Stokes equation together with the heat flow equation. His calculations suggest that the melt pool rotates approximately five times before solidifying. The marker experiments of Takeda (32), also indicate that a very rapid mixing takes place within the melt pool and one of great complexity due to micro eddies, which are difficult to model.

6.4. Laser Surface Alloying

Surface alloying with a laser is similar to laser surface melting except that another material is injected into the melt pool. Laser surface alloying is also similar to surface cladding in that if the cladding process is performed with excess power then surface alloying would result. It is therefore one extreme of surface cladding. The main characteristics of the process are as follows:

* The alloyed region shows a fine microstructure with nearly homogeneous mixing throughout the melt region. Inhomogeneities are only seen in very fast melt tracks (~ 0.5 m/s).
* Most materials can be alloyed into most substrates. The high quench rate ensures that segregation is minimal (33). Some surface alloys can only be prepared via a rapid surface quench, e.g. Fe-Cr-C-Mn (33).
* The thickness of the treated zone can be from 1-2000 μm. Very thin, very fast quenched alloy regions can be made using Q-switched Nd-YAG lasers.
* Some loss of the more volatile components can be expected (34).
* Other characteristics are as for laser surface melting.

Thermochemical diffusion treatments

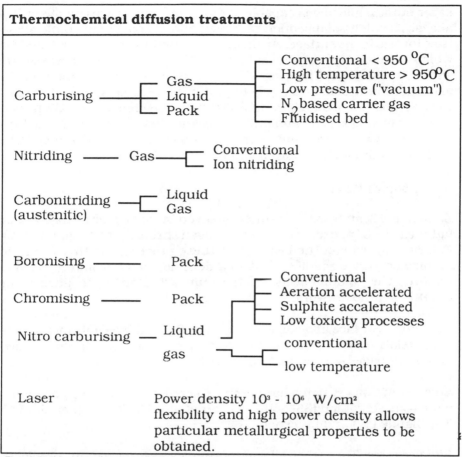

Fig. 6.24. Competing surface alloying methods.

<u>6.4.1. Process Variations</u>

The variations in processing are similar to those for surface melting except that an alloy ingredient has to be added. The alloy can be placed in the melt zone by:

1.   Electroplating (35).
2.   Vacuum evaporation.
3.   Preplaced powder coating (36).
4.   Thin foil application.
5.   Ion implantation.
6.   Diffusion, e.g. boronising (37).
7.   Powder blowing (38).
8.   Wire feed.
9.   Reactive gas shroud (39), e.g. C_2H_2 in Ar or just N_2.

Laser surface alloying is capable of producing a wide variety of surface alloys. The high solidification rate even allows some metastable alloys to be formed in the surface. All this can be done by a non-contact method which is relatively easy to automate. The competing processes are shown in Fig. 6.24. The laser offers precision in the placement of the alloy, good adhesion and vastly improved processing speeds. Provided the speed is lower than a certain figure (e.g. 70mm/s for 2kW power) then the mixing is good and uniform. Some alloys suffer from cracking and porosity which may put restrictions on shrouding and preheat. The surface profile can be quite smooth with a small ripple of around 10μm.

6.4.2. Applications

Titanium: Ti can be readily surface alloyed by carbon or nitrogen. The latter can be supplied by having a nitrogen shroud gas (Bergmann (12), Folkes (40)). One of the beauties of these processes is that the hard carbide or nitride solidifies first as a dendrite which would be hard to remove. The colour effects on titanium are starting to attract the attention of the art world.

Cast iron: Surface alloying with Cr, Si or C are all possible methods to make relatively cheap cast irons into superficially exotic irons. A study has been made by Zhen da Chen (21).

Steel: Numerous systems have been explored. Cr by melting chromium plate (Christodoulou (35)); Mo (Tucker et al (41)); B (Lamb et al. (37)); Ni (Chande (42) and Lumsden (26)).

Stainless steel: The carbon alloying of stainless steel by melting preplaced powder has been studied by Marsden (43).

Aluminium: Surface hardening of aluminium by alloying with Si, C, N and Ni has been shown possible by Walker et al (36) and others.

Superalloys have been alloyed with chromium by Tien et al (44).

Surface alloying has many advantages and great flexibility. Applied by laser the process offers the possibility of surface compositional changes with very little distortion and surface upset. This has thus put engineers in the position that they could have the material they require for the surface and the material they require for the bulk. The problem of the choice is exhausting to contemplate !

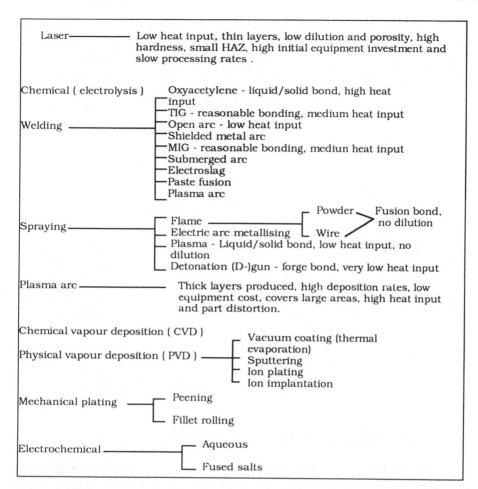

Fig 6.25. Competing cladding techniques.

6.5. Laser Cladding

The aim of most cladding operations is to overlay one metal with another to form a sound interfacial bond or weld without diluting the cladding metal with substrate material. In this situation dilution is generally considered to be contamination of the cladding which degrades its mechanical or corrosion resistant properties. There are many cladding processes as shown in Fig.6.25. Thick section cladding (> 0.25 mm) is frequently carried out by welding methods; substantial melting of the substrate is produced and therefore dilution can be a major problem. Dilution is observed in tungsten inert gas (TIG), oxy-acetylene flame or plasma surface welding processes in which the melt pool is well stirred by electromagnetic, Marangoni and convective forces. This dilution necessitates laying down thicker clad layers to achieve the required clad property, but does have the advantage of a good interfacial bond. Negligible dilution is achieved in other cladding processes which rely on either forge bonding or diffusion bonding; forge bonds are made through the impact of high speed particles with the substrate (e.g. D-gun) or clad layer and diffusion bonding occurs between a solid and liquid phase. The fusion bond is usually the strongest and most resistant to thermal and mechanical shock, provided brittle intermetallics are not formed. A comparative study of the dilution, distortion, wear and other properties of clad layers made with a laser, plasma, vacuum furnace, TIG or oxy acetylene flame has been made by Monson (45,46).

Among the laser cladding routes are those which melt preplaced powder (47), or blown powder (38), those which decomposed vapour by pyrolysis (48), or photolysis (49) as in Laser Chemical Vapour Deposition, (LCVD), those which are based upon local vaporisation as in Laser Physical Vapour Deposition (LPVD) or sputtering and those based on enhanced electroplating or cementation (50). These latter three processes are discussed separately in Sections 6.7, 6.8 and 6.9.

The two most common methods of supplying the cladding material are
i)   Preplacement of cladding material powder on the substrate.
ii) Inert gas propulsion of material powder into a laser generated molten pool.

6.5.1. Laser Cladding with Preplaced Powder

Cladding with preplaced powder is the simplest method provided the powder can be made to stick until melted, even while the area is being shrouded in inert gas. Some form of binder is usually used. The pre-placed powder method involves scanning a defocussed or rastered laser beam over a powder bed, which is consequently melted and welded to the underlying substrate. Minimal dilution effects were observed for a wide range of processing parameters. Theoretical modelling of movement in the molten front (51) has shown that the melt progresses relatively swiftly through the thermally isolated powder bed until it reaches the interface with the substrate. At this point the thermal load increases due to the good thermal contact with the high thermal conductivity substrate, causing resolidification. The results of the model are shown in Fig. 6.26. This figure illustrates why a large operating region for achieving low dilution exists; but it also shows that only a small part of this region gives a fusion bond. In fact it would be very difficult to achieve a low dilution fusion bond by this method.

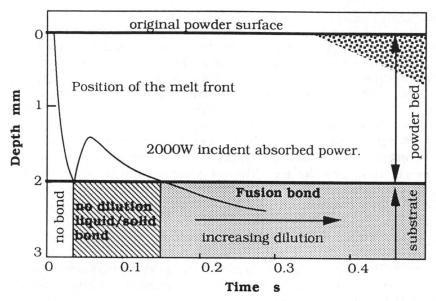

Fig. 6.26. Theoretical calculation of the position of the melt front during preplaced powder cladding (51).

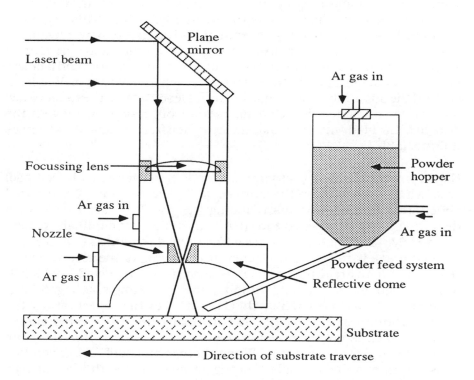

Fig. 6.27. Arrangement for laser cladding by the blown powder technique.

6.5.2. Blown Powder Laser Cladding (38)

The main interest in this process is because it is one of the few cladding techniques which has a well defined heated region, a fusion bond with low dilution and is adaptable to automatic processing. The arrangement for cladding by blown powder is illustrated in Fig. 6.27. In this figure an optical feedback system is shown. A reflective dome such as this has been shown by Weerasinghe (38) to recover around 40% of the delivered power. This is necessary when cladding surfaces of variable reflectivity such as machined and shot blasted surfaces as shown in Fig. 6.28. Blown powder cladding has the low dilution associated with forge bonded processes but the good surface strength and low porosity associated with the welding processes. The covering rate for laser powers greater than 5 kW is attractive (Fig. 6.29.a,b) and when consideration of powder costs and after machining costs are taken into account then the process becomes economically comparable with other processes for covering large areas. This process has the ability to cover very small areas and in particular areas near to thin walls which might be thermally sensitive, since there are no associated hot gas jets.

There is a reasonably large operating window for successful low dilution fusion bonded cladding as illustrated in Fig. 6.30. This region is bounded by a dilution limit - dependent on the excess power available after the powder has been melted; an aspect ratio limit - dependent upon whether runs can be overlapped without interrun porosity; and a power limit - dependent upon whether there is sufficient power to melt at least some of the substrate (52). It has been suggested that the extensive low dilution region is due to the solidification front rising swiftly with the growth of the clad, meaning that the interface freezes almost as soon as it forms (53).

Blown powder cladding is essentially conducted over a small melt pool area which travels over the surface of the substrate. The thermal penetration is minimal thus reducing the problems associated with distortion and heat affected zone (HAZ), although not eliminating them completely. The distortion varies with the clad thickness (Fig. 6.31). If area coverage is to be achieved then overlapping of the clad tracks is required. There are three basic cross-sections of such clad tracks (52) (Fig. 6.32.a,b,c). For cladding without porosity the angle, α (Fig. 6.32.c) must be acute, as in Fig. 6.32a; this is defined by the aspect ratio of the track. Dilution in Fig. 6.32.b is principally due to extreme energy delivery which is not absorbed in melting the powder. It is therefore defined by an energy balance. When there is a lack of fusion bonding a discontinuous clad or 'balling' of the powder results; this will also be determined from an energy balance.

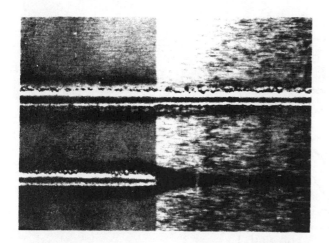

Fig. 6.28. The left side of the substrate was shot blasted. The right side had a ground finish. The upper track was made without the reflective dome; the lower track was made using the dome to recycle reflected energy (38).

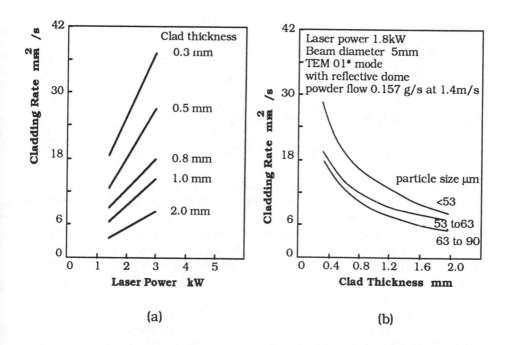

(a)   (b)

Fig. 6.29. The variation of typical laser cladding rates with power, clad thickness, and powder particle size using the blown powder process with the reflective dome (38).

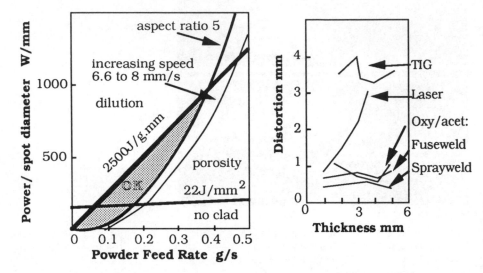

Fig. 6.30. The operating window for blown powder laser cladding.

Fig. 6.31. Hardfacing thickness versus the distortion of a standard size sample for coatings produced by various techniques

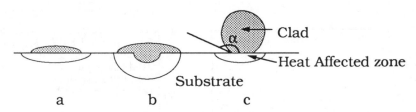

Fig. 6.32. The three basic cross section profiles for single track clad beads.

6.6. Particle Injection (54,55)

This process is similar to laser cladding by the blown powder route except that the particles blown or projected into the laser melt pool do not entirely melt (56); this creates a structure similar to a Macadam road (Fig.6.33). The main advantages for this process are improved hardness and wear resistance with reduced friction coefficients in some systems. Process variations centre around the particle delivery system, delivery pressure (vacuum or atmospheric) and the gas shrouding systems.

In order to achieve a good surface layer the hard embedded particles must be wetted by the metal matrix and they must have strong bonding to it. It is also desirable that the particles do not suffer too much dissolution while lying in the melt pool. These requirements mean the particle and the surface must be clean and the level of superheat must be kept as low as possible, compatible with the wetting condition.

This process is still at the laboratory stage but shows considerable promise in hardening aluminium and its alloys by the injection of TiC, SiC, WC or Al_2O_3 particles. It has also been applied to stainless steel.

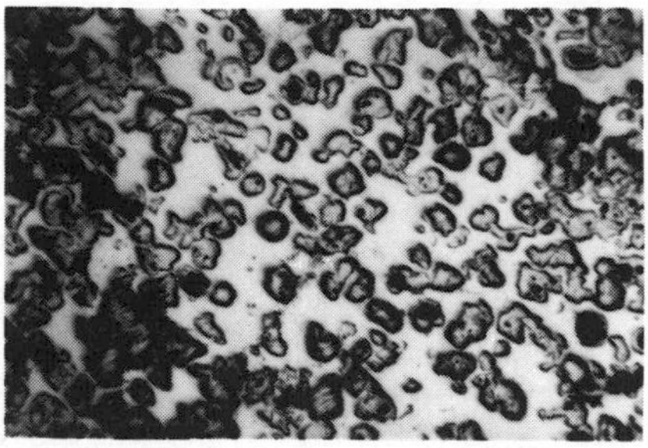

Fig. 6.33. Macro graph showing a laser clad made with mixed powder of stellite and TiC. The TiC has not melted and forms a type of "Macadam" road. x 50

6.7. Surface Texturing (57)

A chopped laser beam is used to make a regular pattern of small pits or dimples in the surface of a texturing roll in a temper mill. Sheets passed through the mill will be surface dulled to aid paint adhesion and flow as well as improving press formability and work handling, Fig. 6.34. Shot blasting or electro-spark machining are usually used to roughen the roll surface. However, these alternative processes have a random roughness which results in long wavelength undulations and imperfect spread of paint. The laser dulled sheets do not have this long wavelength effect to the same extent and so give almost perfectly flat paint surfaces. This may allow a reduction in the number of paint coats required or simply give a higher quality product for no increase in cost. Fig. 6.35 illustrates the long wavelength and roughness characteristics of sheets treated conventionally and by the laser texturing method.

6.8. Enhanced Electroplating

The irradiation by a laser beam of a substrate used as a cathode during electrolysis, causes a drastic modification of the electro-deposition process in the irradiated region. Interesting aspects of the process are:
* The possibility of rapid maskless patterning.
* The possibility of enhanced plating rate on selected areas.
* The possibility of modifying the structure of electrodeposited coatings.
The first attempt to use the laser in combination with electroplating is thought - possibly mythically - to have been done by someone interested in using a Laser Doppler Anemometer (LDA) to measure the flow at the cathode during plating. Instead of measuring flow rate he found a new process! Possibly this story belongs to IBM-Thomas J.Watson Research Centre, since Von Gutfeld gives them the credit for the invention (58). From these experiments it was shown that the plating rate increased in the irradiated zone by up to 1000 times!

Shot blasting roll surface.   Laser machining roll surface.

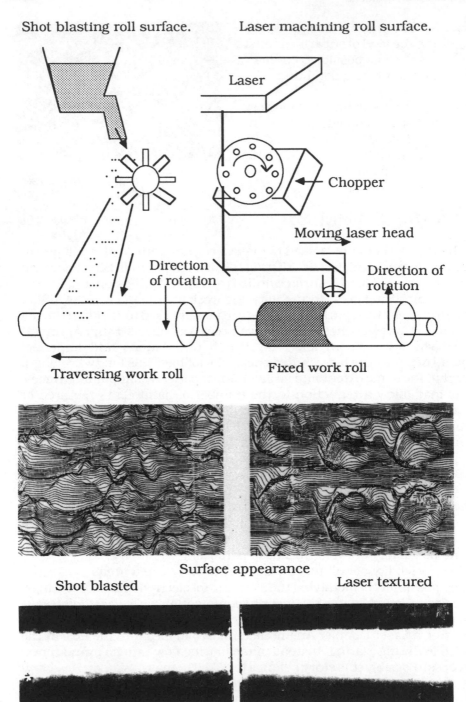

Surface appearance

Shot blasted   Laser textured

Appearance of a strip light reflected in a painted surface

Fig.6.34. Laser texturing of rolls for making textured rolled steel sheets (57).

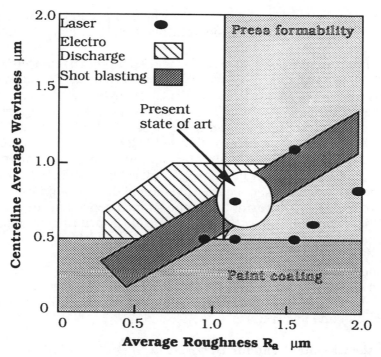

Fig. 6.35. Plot of surface roughness versus surface waviness for a painted surface of press steel (from57).

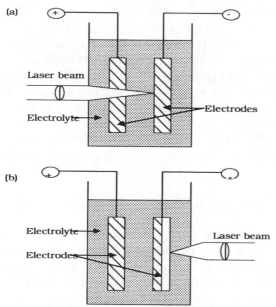

Fig. 6.36. Schematic representation of (a) directly irradiated cathode and (b) cathode irradiated from the back.

Three processes have emerged: laser enhanced plating, laser induced electroless deposition or immersion plating, and laser assisted etching.

Laser enhanced electroplating of Ni, Cu and Au has been reported by Von Gutfeld. Power densities of 10 to 10^4 W/mm^2 are used. The laser used depends upon the transmissivity of the electrolyte and the absorption properties of the cathode. Thus Argon ion lasers (λ=514.5nm) are used for copper and nickel solutions while Kr$^+$ lasers (λ=647.1nm) or Nd:YAG lasers (λ=1060nm) are used for yellow gold electrolytes. The power can be directed onto the front of the cathode or through a transparent cathode (Fig. 6.36). A light pen has been developed by IBM in which a jet of electrolyte with a laser beam waveguided down it impinges on the area to be treated. The result is a dome shaped deposit around 0.5mm wide. The potential applications are maskless printing or repair of circuit boards.

Laser induced electroless deposition has been tried for a number of systems e.g. copper from copper sulphate/hydrochloric acid (Al-Sufi et al (59)). This process poses the possibility of printing circuits on non conductors such as ceramics.

6.9. Laser Chemical Vapour Deposition

Blowing thermally sensitive vapour onto a laser generated hot spot can cause a deposit to be formed by pyrolysis. The rate of deposition is controlled by chemical reaction rates (Arrhenius Equation) up to certain deposition rates dependent on the surface temperature; above these temperatures the process is controlled by mass transport. Under mass transfer control the quality of the deposit falls markedly from a smooth sheet to a rough surface and ultimately to powder. The rates are illustrated in Fig. 6.37 from (60). The rate of deposition is usually slow being a few μm/minute for most processes due to the need to avoid mass transfer control. Alternatively the vapour could be directly broken by the photons in a process of photolysis. This is particularly relevant to processing with the excimer laser. Perez-Amor (61,62) has deposited SiO$_2$ in this manner.

6.10. Laser Physical Vapour Deposition

The laser beam can be directed onto a target situated in a vacuum chamber. The target evaporates and the vapour condenses on the substrate among other areas. This process has the advantage of extreme cleanliness in the heating technique. It is possible using an excimer pulse to ablate the surface of the target resulting in a deposit on the substrate which is of the same composition as the target material

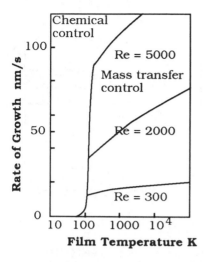

Fig. 6.37. Temperature dependence of the growth rate of CoO deposit during Laser Chemical Vapour Deposition (LCVD) of co-balt acetylacetonate from a jet of concentration 0.3ppm (60).

with no difference due to vapour pressures. Such a process has been described for the deposition of superconducting alloys (63).

6.11. Non Contact Bending (64)

Non contact bending is included under surface treatments since the process involves the generation of surface stress. In this process, metal and alloy sheets are irradiated with a defocussed CO_2 laser beam scanning at speeds up to 15 m/min (250 mm/s). The sheet can be bent into a "V" or "C" or other shape by repetitively scanning the beam over the sheet at particular zones.   The final bending angle of the irradiated sheet was almost in proportion to the number of laser beam scans, Fig. 6.38.   This bending action is caused by thermal stresses in the sheet due to the extremely rapid heating and cooling process during the laser irradiation. The bending angle was affected by the mechanical and thermal properties of the work material as well as the properties of the input laser energy.

The laser energy is of sufficient intensity to cause a steep thermal gradient but insufficient to cause melting. The resultant thermal expansion of the material causes the thermal stress which deforms the material plastically. Only if there is some plastic deformation will there be any permanent bending.

The process is slow, as seen in Fig. 6.38, but a non contact bending process must have possibilities. A variant used in production is to scan a laser along the edge of a rotating rod such that it just misses the rod. If there is any bend in the rod the inside of the bend will strike the laser beam, become heated and expand thus straightening the rod.

6.12. Magnetic Domain Control (65)

In this process the laser induces thermal stress fields which cause the generation of subdomains leading to a significant decrease in eddy

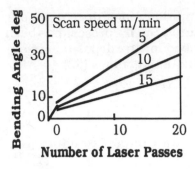

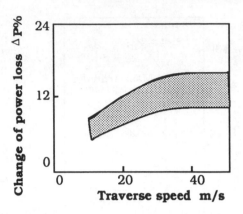

Fig. 6.38. The relationship between the bending angle and the number of laser scans in laser forming. Laser power 1.5kW; spot size 5.4mm; carbon coating; 304 stainless steel.

Fig. 6.39. Reduction in core loss versus laser traverse speed (after 65).

current losses in transformers (Fig. 6.39). Electrical machines and transformers contain metal sheets having particular magnetic properties. These are planned to guide the magnetic field with only minimal losses. High-power transformers exclusively contain grain oriented electrical sheets of an efficiency of up to 99%.

A CW Nd:YAG laser has been used by Neiheisel (66) to refine the magnetic domain size in transformer steel, whilst leaving the insulating coating on the steel intact. In the laser domain refinement process (65), a high power focussed beam is scanned rapidly (100m/s) across the surface of 3% silicon-iron, as used in electrical transformers. The material after treatment shows no visible surface change; however, a reduction in the magnetic core loss has occurred.

A thermal shock is believed to be imparted to the microstructure which causes slip plane dislocations to form, thereby producing new magnetic domain wall boundaries. By adjusting the spacing of the scanned laser lines, the energy lost due to moving the domain walls back and forth under the action of the applied AC field in the transformer is minimised. The laser lines restrict the length of the domains which also acts to control the width of the domains. Thus by adjusting the spacing of the laser lines, the domain sizes can be controlled, i.e. refined. The process is illustrated in Fig. 6.40.

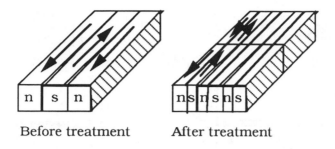

Before treatment   After treatment

Fig. 6.40. The principles of domain refinement.

6.13. Stereolithography (67)

Stereolithography or 3D printing is a process by which intricate models are directly constructed from a CAD package by polymerising a plastic monomer. Ultra-violet light of wavelength 325 nm from a low power He-Cd laser shines onto a pool of liquid monomer causing it to selectively polymerise and set. The process relies on a scanned laser beam to selectively harden successive thin layers of photopolymer, building each layer on top of the previous layer until a three-dimensional part has been constructed. The photopolymer used for stereolithography has a very thick consistency and quickly hardens upon exposure to specific wavelengths of light. The steps required to produce a part by this method involve designing the model on a computer aided design (CAD) system, slicing and preparing the data by computer, creating the model and postcuring.

6.13.1. Model Design

The first step in creating a model by stereolithography is to perform the design function on a 3D CAD system. The CAD image is constructed as either a solid or surface model complete with final wall thicknesses and interior details. Next, the solid model image is orientated on the screen into the position which best facilitates construction by stereolithography. A supporting structure made of thin cross webs may be required for elements of the model which overhang because the stereolithography process constructs the model layer by layer beginning from the bottom surface. The use of support structures may be minimised by optimising the orientation of the part. After the model has been made, the support structure would be broken away or cut off.

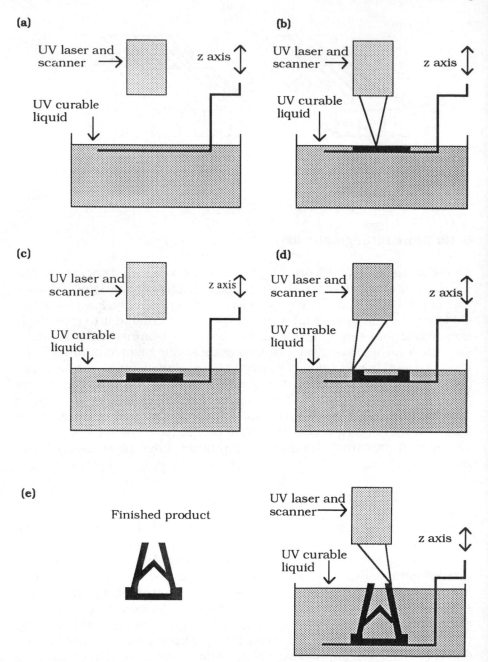

Fig. 6.41. The sequence for making a part by stereolithography (67).
a) The platen is lowered to a preset depth for the slice being cast.
b) The initial layer is polymerised by u/v light to form the bottom layer from the CAD image to be cast.
c) and d) The cycle is repeated for each layer; until the finished part (e).

6.13.2. Model Slicing and Data Preparation

On completion of the design, the electronic model is transferred to the stereolithography computer where it is electronically sliced into thin layers. The thickness of the individual layers is selected by the operator and is generally thinner in areas requiring the greatest accuracy. The computer then breaks each layer into a gridwork of vector data that will later be used to guide the direction and velocity of the laser beam.

6.13.3. Model Creation by Stereolithography

The stereolithography process is illustrated in Fig. 6.41. It begins by lowering a platen to a precise distance into a vat of photopolymer. The distance the platen is lowered corresponds to the thickness of the model layer being hardened. A wiper ensures the plate is uniformly covered and that no waves occur on the monomer surface. Next, the laser beam is activated and guided in the x and y axes by two computer controlled scanning mirrors. The motion of the laser beam selectively hardens the photopolymer in the areas corresponding to the first slice of the model. The first layer solidified becomes the bottom layer of the model. The platen is again lowered into the vat a precise distance and the laser draws the second layer on top of the first one. This process is repeated as many times as is necessary to recreate the entire object layer by layer. When completed, the platen is raised from the vat and the model is ready for removal of the support structure and postcuring. A typical processing time would be around 1hr with a further 2hrs for post curing to make a 3D model in around 3hrs with variations for complexity.

6.13.4. Postcuring

Maximum polymer hardness is not attained directly from the stereolithography process and therefore, some postcuring is usually required. This is accomplished by exposing the model to high intensity ultraviolet light in a special curing unit. Exposure times vary from thirty minutes to two hours depending on part geometry.

Postcuring is also used as a method of quickly hardening thick walls and large volumes in models. Because polymerising large volumes of material with the low power laser is very time consuming and inefficient, large volumes are more quickly created by entrapping liquid polymer within the model structure to be hardened later during the postcuring cycle.

The capabilities of stereolithography are determined by the power of the laser, the accuracy of the scanners and optical system and the type of photopolymer being used. Most currently identified applications of stereolithography take advantage of the ability to directly interface with electronic data bases and create complex models without machining. Important applications which have been proposed or demonstrated include:

* Three-dimensional topographic maps from satellite data.
* Architectural models complete with complex internal structures.
* Templates and cams for mechanical controls.
* Three-dimensional plots of scientific and engineering data.
* Manufacturing-related models.

6.14. Paint Stripping

It is possible to volatilise paint from a metal surface by laser irradiation without damaging the underlying metal. United Technology is currently designing a system (Automated Laser Paint Stripping (ALPS) (68)) which can remove paint from fighter aircraft or helicopters at the rate of $1m^2/$ min using a pulsed 6kW CO_2 laser. The pulse energy is 6J per pulse the pulse lasts for 30µs at a repetition rate of 1kHz. This technique represents a significant advance over chemical stripping alternatives; in particular it avoids problems associated with the corrosive nature of chemical strippers. The reduction in waste by a factor of around 100 is a further attraction.

6.15. Conclusions

Optical energy is an ideal form of energy for surface treatment. The uses of laser surface treatment include surface heating, bending, melting, alloying, cladding, texturing and stereolithography. The advantages offered by the laser are the highly localised clean nature of the process, low distortion and high quality of finish. It is thus not surprising to find that laser processing of surfaces is a subject currently enjoying much research and industrial interest and with the development of highly automated workstations and lasers which are more powerful, reliable and compact, it is now economically and practically feasible to use lasers industrially for surface treatment processes.

References

1. V.G.Gregson, Laser Heat Treatment in Laser Materials Process-
    ing ed. M.Bass, North-Holland Publishing Company, 1984,
    Chapter 4.
2. T.Bell, Surface Heat Treatment of Steel to Combat Wear,
    Metallurgia March 1982 49 (3) 103-111.
3. F.Dausinger, M.Beck, T.Rudlaff and T.Wahl, On Coupling
    Mechanisms in Laser Processes, Proc. LIM 5, ed. H.Hugel, Publ.
    IFS Ltd., 1988 pp177-186.
4. W.M.Steen and C.Courtney, Laser Surface Treatment of En8
    Steel using a 2 kW CO_2 Laser, Metal Tech. 1979 6 (12) 456.
5. O.Kubaschewski, Iron-Binary Phase Diagrams, Publ. Springer-
    Verlag, 1982.
6. M.Atkins, Atlas of Continuous Cooling Transformation Diagrams
    for Engineering Steels, Publ. British Steel Corporation.
7. M.Sharp and W.M.Steen, Investigating Process Parameters for
    Laser Transformation Hardening, Proc. 1st. Int. Conf. on Surface
    Engineering, paper 31 publ. Welding Institute, Cambridge 1985.
8. J. Mazumder and W.M.Steen, J. App. Phys. 51 (3) 1980 pp941.
9. M.F.Ashby, K.E.Easterling, The Transformation Hardening of
    Steel Surfaces by Laser Beam, Acta Metall., 32 1935-1948
    (1984).
10. C.Chan, J.Mazumder and M.M.Chen, A Two Dimension Tran
    sient Model for Convection in a Laser Melted Pool, Met. Trans. A
    vol. 15A Dec. 1984 pp2175-2184.
11. M.F.Ashby and H.R.Shercliff, Master Plots for Predicting the
    Case-Depth in Laser Surface Treatments, Engineering Dept.,
    Cambridge University , October 1986. Document CUED/C-Mat/
    TR 134.
12. H.W.Bergmann, Laser Surface Melting of Iron-base Alloys, Proc.
    NATO Advanced Study Institute on Laser Surface Treatment of
    Metals, San Miniato, Italy, Sept. 2-13, 1985, pp 351-368.
13. Metals Handbook, 9th ed., Vol. 1 Properties and Selection: Irons
    and Steels, Publ. ASM, Metals Park, Ohio 44073, 1978.
14. Lui Zhu, Ph.D. thesis, Liverpool University, 1991.
15. J.Mazumder, Laser Heat Treatment: The State of the Art, J. Met.,
    May 1983 35 (5) 18-26.
16. M.A.Bramson, Infra-red Radiation: a Handbook for Applications,
    Plenum Press, New York, 1968.
17. F.O.Olsen, Laser Material Processing at the Technical University
    of Denmark, Proc. Materialbearbeitung mit CO-Hochleist, Stutt
    gart April 1982 paper 2.
18. M.Jorgensen, Met. Costr. Feb. 1980 12 (2) 88 1980.

19. A.V.La Rocca, Laser Applications in Manufacturing, Scientific American March 1982 pp80-87.
20. V.M.Weerasinghe and W.M.Steen, Laser Cladding by Powder Injection, Proc. Conf. Lasers in Manufacturing 1, publ. IFS publ. Ltd., Kempston Bedford ed. M. Kimmitt Nov. 1983 pp125-132.
21. W.M.Steen, Z.D.Chen and D.R.F.West, Laser Surface Melting of Cast Irons and Alloy Cast Irons, Industrial Laser Annual Hand book 1987, ed. David Belforte and Morris Levitt pp80-96.
22. I.C.Hawkes, W.M.Steen and D.R.F.West, "Laser Surface Harden ing of S G Cast Iron", Metallurgia 1983 50 (2) 68-73.
23. D.N.H.Trafford, T.Bell, J.H.P.C.Megaw and A.D.Bransden, Laser Treatment of Grey Iron, Metals Technology 1983 10 (2) 69-77.
24. I.C.Hawkes, A.M.Walker, W.M.Steen and D.R.F.West, Applica tion of Laser Surface Melting and Alloying to Alloys Based on the Fe-C system, Lasers in Metallurgy 2, Los Angeles Feb. 1984, ed. K.Mukherjee and J.Mazumder ASM publ. 1984.
25. M.Lamb, W.M.Steen and D.R.F.West, Structure and Residual Stresses in Two Laser Surface Melted Stainless Steels, Proc. Conf. Stainless Steel '84, Gothenburg Sweden Sept. 1984.
26. J.B.Lumsden, D.S.Gnanamuthu and R.J.Moores, Fundamental Aspects of Corrosion Protection by Surface Modification ed. E.McCafferty, C.R.Clayton and J.Oudar, publ. Electrochem. Soc. Pennington, NJ 1984 pp122.
27. T.R.Anthony and H.E.Cline, J. App. Phys., 48, 9 (1977).
28. J.A.Folkes, Ph.D thesis, University of London, 1985.
29. J.A.Folkes, P.Henry, K.Lipscombe, W.M.Steen and D.R.F.West, Laser Surface Melting and Alloying of Titanium Alloys, Proc. 5th. Int. Conf. on Titanium, Munich Sept. 1984.
30. M.C.Flemings,Solidification Processing,McGraw Hill Book Co., 1974.
31. I.C.Hawkes, M.Lamb, W.M.Steen and D.R.F. West, Surface Topography and Fluid Flow in Laser Surface Melting, Proc. CISFFEL Lyon France Sept. 1983 vol.1 pp 125-132 publ. Le Commissariat a l'Energie Atomique, France.
32. T.Takeda, W.M.Steen, D.R.F.West "Laser Cladding with Mixed Powder Feed", Proc. conf. ICALEO '84 Boston Nov. 1984 pp151-158.
33. J.Mazumder, Wear Properties of Laser Alloyed Fe-Cr-Mn-C Alloys, Proc. ICALEO '84, Boston Nov.1984.
34. A.Blake and J.Mazumder, Control of Mg loss during Laser Weld ing of Al-5083 using a Plasma Suppression Technique, ASME J. of Eng. for Ind. Aug. 1985.
35. G.Christodoulou and W.M.Steen, Laser Surface Treatment of Chromium Electroplate on Medium Carbon Steel, Proc. 4th Int. Conf. on Laser Processing, Los Angeles Jan. 1983.

36. A.M.Walker, D.R.F.West and W.M.Steen, Laser Surface Alloying of Ferrous Materials with Carbon, Proc. Laser '83 Optoelectronik Conf., ed. W.Waidelich, Munich June 1983 pp322-326.

37. M.Lamb, C.Man, W.M.Steen and D.R.F.West, The Properties of Laser Surface Melted Stainless Steel and Boronised Mild Steel, Proc. CISFFEL Lyon Sept. 1983 pp 227-234 Publ. Le Commissariat a L'Energie Atomique, France.

38. V.M.Weerasinghe and W.M.Steen, Laser Cladding with Pneumatic Powder Delivery, Proc. 4th Int. Conf. on Lasers in Materials Processing, Los Angeles, Jan. 1983, ed. E.A.Metzbower. Publ. ASM, Ohio, USA pp166-175, 1984.

39. A.M.Walker, J.Folkes, W.M.Steen and D.R.F.West, The Laser Surface Alloying of Titanium Substrates with Carbon and Nitrogen, Surface Engineering 1985: 1 (1).

40. J.Folkes, D.R.F.West and W.M.Steen, Laser Surface Melting and Alloying of Titanium, Proc. NATO Advanced Study Institute on Laser Surface Treatment of Metals, San Miniato, Italy, Sept. 2-13, 1985, pp451-460.

41. T.R.Tucker, A.H.Clauer, S.L.Ream and C.T.Walkers, "Rapidly Solidified Microstructures in Surface Layers of Laser Alloyed Molybdenum on Fe-C Substrates" Proc. Conf. Rapidly Solididified Amorphous and Crystaline Alloys, Boston Mass. Nov. 1981, pp 541-545 Publ. Elsevier Science Publ. Co. Inc. New York 1982.

42. T.Chande and J.Mazumder, J. App. Phys. 57, 2226 (1985).

43. C.Marsden, D.R.F.West and W.M.Steen, Laser Surface Alloying of Stainless Steel with Carbon, Proc. NATO Advanced Study Institute on Laser Surface Treatment of Metals, San Miniato, Italy, Sept. 2-13, 1985, pp461-474.

44. J.K.Tien, J.M.Sanchez and R.T.Jarrett, "Outlook for Conservation of Chromium in Superalloys", Proc. Tech Aspects of Critical Materials use by the Steel Industry vol. 11-B, Nashville Tenn. USA 4 - 7 Oct. 1982 p 30. Publ. Nat. Bur. Stds., Washington USA 1983.

45. P.J.E.Monson and W.M.Steen, A Comparison of Laser Hardfacing with Conventional Processes, Surface Engineering 1990 vol 6 no 3pp185-194.

46. P.J.E.Monson, Ph.D thesis, University of London 1988.

47. J.Powell, Proc. Conf. on Surface Engineering with Lasers, London, May 1985, Paper 17, publ. Metal Society, London.

48. W.M.Steen, Surface Coating with a Laser, Proc. Conf. Advances in Coating Techniques, publ. W.I. Cambridge, UK (1978), pp175-187.

49. J.Jardieu de Maleissye, Laser Induced Decomposition of Molecules Related to Photochemical Decomposition, Laser Surface Treatment of Metals, ed. C.W. Draper, P. Mazzoldi, Proc. NATO ASI., San Miniato, Italy, 1986. Publ. Martinus Nijhoff, Dordrecht, Netherlands, pp555- 566 1986.

50. J.R.Roos, J.P.Celis and W.Van Vooren, Combined use of Laser Irradiation and Electroplating, Laser Surface Treatment of Metals, ed. C.W. Draper, P. Mazzoldi, Proc. NATO ASI.,ibid. pp577-590 1986.

51. J.Powell, P.S. Henry and W.M. Steen, Laser Cladding with Preplaced Powder: Analysis of Thermal Cycling and Dilution Effects, Surface Engineering, 1988, vol. 4, no. 2 pp141-149.

52. W.M.Steen, V.M.Weerasinghe and P.J.E.Monson, Some Aspects of the Formation of Laser Clad Tracks, Proc. SPIE. Conf. Innsbruck, Austria, April 1986, publ. by SPIE., P.O. Box 10, Bellingham, Washington, 98227-0010 USA, vol. 650 pp226-234.

53. W.M.Steen, Laser Surface Cladding, invited paper, proc. Indo-US workshop on Principles of Solidification and Materials Processing, SOLPROS, Hyderabad, India, Jan. 1988. Publ. ONR., AIBS., 1988.

54. J.D.Ayers et al J. Metals Aug. 1981.

55. J.E.Flinkfeldt Optics and Laser Tech. 1988.

56. J.D.Ayers, Modification of Metal Surfaces by Laser Melt Particle Injection Process, Thin Solid Films 1980.

57. L.G.Hector, S.Sheu "Focussed energy beam work roll surface texturing science and technology" Journ. Mat. Proc. and Manf. Science Vol.2 July 1993 pp63-117.

58. R.J.Von Gutfeld, E.E.Tynan, R.L.Melcher and S.E.Blum, App. Phys. Lett. 35 651-653 (1979).

59. A.K.Al-Sufi, H.J.Eichler and J.Salk, J.App.Phys. 54, 3629-3631 (1983).

60. W.M.Steen, Ph.D. thesis, University of London, 1976.

61. B.Leon, A.Klumpp, M.Perez-Amor, and H.Sigmund, Excimer Laser Deposition of Silica Films- A comparison between two methods, Proc. E-MRS Conf., Strasbourg, 1990. J. App. Surface Sci.1991.

62. T.Szorenyi, P.Gonzales, D.Fernandez, J.Pou, B.Leon, M.Perez-Amor, Gas Mixture Dependency of the LCVD of Silica Films using an ArF Laser, Proc. E-MRS Conf., Strasbourg, 1990. To be published in J. App. Surface Sci. 1991.

63. Ding.M.Q., Rees.J.A., Steen.W.M. "Plasma assisted laser evapora tion of superconducting YBaCuO thin films" 7th Int Conf IPAT 1989 Geneva June 1989.

64. Y.Namba, Laser Forming of Metals and Alloys, Proceedings of Conf. LAMP '87, Osaka (May, 1987) pp601-606.

65. A.Gillner, K.Wissenbach, E.Beyer, G.Vitr, Reducing Core Loss of High Grain Oriented Electrical Steel by Laser Scribing, Proc. 5th. Int. Conf. Lasers in Manufacturing (LIM 5) ed. H. Hugel, Stuttgart, Sept. 1988, publ. IFS (publ.) Ltd. UK pp137-144.
66. G.L.Neiheisel, Laser Magnetic Domain Refinement, LIA vol. 44 ICALEO'84 Boston Nov (1984) pp102-111.
67. Gary F.Benedict, Stereolithography- The New Design Tool for the 1990s Proc 6th Int conf. on Lasers in Manufacturing (LIM6) B'ham, UK pp249-261.
68. The Laser Edge, Publ. UTRC , July 1990.

"You said it was a fast process so do this lot by 12 o'clock!"

Chapter 7

Laser Automation and In-Process Sensing

"Everything must be like something, so what is this like?"
E.M.Forster 1879-1970 Abinger Harvest (1936).

"To govern is to make choices"
Duc de Levis 1764-1830 Politique "Maximes de Politique"xix.

7.1. Automation Principles

The recent developments in industry, particularly through the activities of the Ford Motor Co., where the word "automation" was first used in the 1940s, have sketched a progression through "mechanisation" - the use of machines which enhanced speed, force or reach, but where the control was human; to "automatic" machinery - in which the machine will go through its programmed movements without human intervention and the machine is self regulating; until today we have "automation" - in which there is usually a sequence of machines all controlling themselves under some overall control. In the future there is the prospect of "adaptive control" or "intelligent" machines - in which the machine can be set a task and it teaches itself to do the task better and better according to some preset criteria. The drive towards automation is powered by the possibility of cost reductions, increased productivity, increased accuracy, saving of labour, greater production reliability, longer production hours, better working conditions for the human staff, increased flexibility of production to meet the needs of changing markets and improved quality. This list is a formidable argument for automation but it is only justified for certain production volumes. Fig. 7.1 gives an idea of the stages which are most economical in setting up an automatic production facility. If very few pieces are needed then it is cheapest to

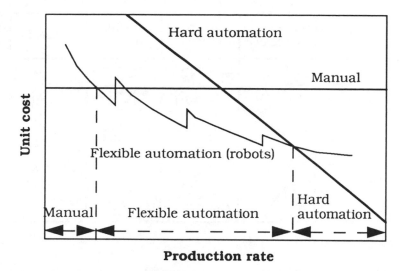

Fig. 7.1. Variation of unit costs with production rate for different manufacturing strategies.

make them by hand. If a very large number of pieces are needed then it is cheapest to make them on a purpose built production line - "hard automation". In between there is the relatively new area of flexible manufacturing, possibly using robots and linked machines. This middle zone in production size is growing due to the manufacturing market becoming more fashion conscious and pandering to the human appetite for novelty and change. Into this manufacturing pattern the laser has a place as another tool, but one with some significant advantages:

Firstly, it is very **flexible** in the way it can be programmed to direct the optical energy. Fig. 7.2 illustrates some of the options. One of the most flexible forms of laser beam guidance is via a robotic beam delivery system. However the accuracy and neatness of the laser, particularly as a cutting tool, shows up the poor line following capability of current robots and this low level of accuracy in the robot precludes its use for many applications. Fig. 7.3 gives an idea of the growth of applications which would result from a successful development of an accurate robot. There is surprisingly little effort being devoted to considering robot control algorithms for the drive mechanisms. Some work is discussed in (1) amongst others. The main effort in robotic research seems to be spent on controlling the sequencing of robots rather than their accuracy.

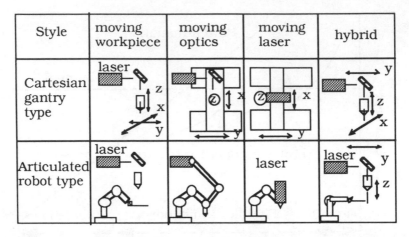

Style	moving workpiece	moving optics	moving laser	hybrid
Cartesian gantry type	laser			y
Articulated robot type	laser		laser	laser y z

Fig. 7.2 Examples of automatic laser workstations.

Secondly, there is very little environmental disturbance in delivering optical energy. For example, there is no electric field, no magnetic field, (except in the electromagnetic radiation itself), there is no sound, no light or other optical signal (except at the frequency of the laser beam), there is no heat, or mechanical stress. Thus any signal in these areas will probably have come from the process itself. This gives a wide **open window for in-process diagnostics**, which is unique to the laser.

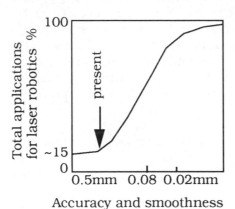

Accuracy and smoothness

Fig. 7.3 Market for laser robotics vs accuracy

In order to have a self regulating system for a laser or any machine it will have either an open or closed loop controller. The open loop controller might be a clock if the sequencing is done by time rather than events. In this case the machine actions are taken automatically but without reference to the state of the process or the position of the machine. If it is controlled by events then it would be a closed loop control system. A schematic of a closed loop controller is shown in Fig. 7.4. It can be seen that the control sequence is:

1. A process variable or product quality is measured.
2. The signal is compared to the desired value and an error detected.
3. This error initiates a change in the process manipulators or drives, thus affecting the process.
4. What is changed and by how much is decided by the controller.

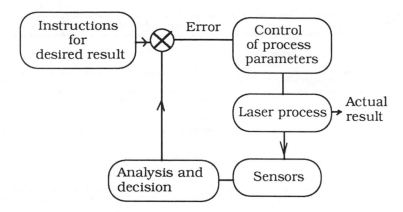

Fig. 7.4 Diagram of the structure of a closed looped control circuit.

The start - if not the heart - of this process is to be able to take the signal from the process while the process is running and to do so sufficiently quickly that the error detection can be made and the machine corrected before there is any considerable product waste. In-process signalling should be and is fast becoming one of the strengths of the laser. This we will discuss next and finally, in Sections 7.3 and 7.4 we will discuss the possibility of in-process control and "intelligent" control systems.

7.2. In-Process Monitoring

For the control or monitoring of laser material processing, the in-process signals for the variables listed in Table 7.1 are required. Many ideas are being and have been devised for these tasks. Table 7.2. lists some of the main concepts being investigated or which have been engineered. This is not a complete list, but nearly so. It does, I hope, illustrate the wide number of sensing options open for laser process monitoring. Some of these techniques we will now discuss in more detail.

Table 7.1	Principle variables which characterise a laser process.	
Beam	**Workstation**	**Workpiece**
Power	Traverse speed	Surface absorptivity
Diameter	Vibration, stability	Seam location
Mode structure	Focal position	Temperature
Location	Shroud velocity and direction	"Quality" of product

Table 7.2	Some in-process sensors currently under investigation or on the market.			
Signal	Sensor	Commercial or Research	Sensor	Commercial or Research
Beam power	Laser Beam Analyser (LBA)	Commercial	Leakage from cavity mirror	Commercial
Beam diameter	LBA	Commercial	Perforated mirror	Commercial
and mode	Hollow needle	Commercial		
Location	Acoustic mirror	Research	Modified LBA	Research
	LBA	Commercial	Scanning slot/beam splitter	Research
	Edge thermocouples	Research		
Traverse speed and	Encoders	Commercial	Linear Moire encoders	Commercial
table position	Tachometers	Commercial	Laser interferometer	Commercial
	Laser doppler anenometer(LDA)	Commercial		
Vibration/stability	Accelerometers	Commercial	LDA	Commercial
	Strain guages	Commercial		
Focal position	Infra red	Research	Pressure	Research
	Capacitance	Commercial	Feeler	Commercial
	Inductance	Commercial	Laser diode	Research
Shroud gas	Nozzle pressure	Commercial	Speckle interferometry	Research
Velocity	Schlieren	Research		
Surface absorption	Acoustic mirror	Research	Back reflection	Research
Seam location	Optical	Res/com	Acoustic	Research
	Pressure	Research		
Cutting quality	TV camera on spark discharge	Research	Acoustic mirror	Research
	Temperature of cut face	Research	Viewing down beam	Research
Welding quality	Acoustic mirror	Research	Plasma charge sensor	Research
	Acoustic workpiece	Research	Laser probe	Research
	Sonic microphone	Research	Acoustic nozzle	Research
	Optical emissions	Research	Video camera	Research
	Electric signals	Research		
Surface hardening	Temperature	Res/com	Acoustic	Research
quality	Infra red	Research		
Cladding dilution	Inductance	Research		
Powder feed rate	Pressure	Research	Vibration	Research
	Stress	Research		

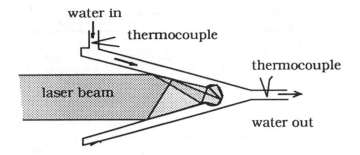

water in

thermocouple

thermocouple

laser beam

water out

Fig. 7.5. Diagram of a standard cone calorimeter power meter/beam dump.

7.2.1. Monitoring Beam Characteristics

The laser power is normally measured when the beam is not being used by some technique which totally blocks the beam. For example, most lasers are fitted with a beam dump which doubles as a calorimeter. A conical beam dump is illustrated in Fig. 7.5. The power is measured as the rise in temperature of the flowing water. An absolute blackbody calorimeter for measuring the power is illustrated in Fig. 7.6. This simple device is highly mobile, but does require a lens to focus the beam into the spherical absorbing chamber, which is usually coated black on the inside. This need for a lens is not usually a disadvantage since the most interesting place to know the power is after the lens at the point where it is to be used. Care must be exercised that all the beam enters the calorimeter. The power can then be read off from a chart recorder reading of the thermocouple output. The equation for the rate of heating of the calorimeter is:

$$P = MC\{dT/dt\}_{heating} - MC\{dT/dt\}_{cooling}.$$

If the calorimeter is made of copper then the specific heat, C, is known to be 4300 J/kgC. The value of the mass of the calorimeter, M, is easily found by weighing. The slopes for heating and cooling come from the chart recorder read out. There is usually some surprise at how much power is lost between the laser and the workpiece by mirrors and lenses (approximately 1-2% /optical element). Such devices are of no use for in-process sensing. It is amazing that some production machines, fitted only with these methods of power sensing, have the laser on for considerable periods of time and therefore have no way of being monitored - they are simply running on faith that the manufacturer has made a stable product. Table 7.3. lists a number of the techniques which have been patented or developed for the in-process monitoring of a laser beam; that is monitoring while the beam is being used. Of these processes the Laser Beam Analyser and the Acoustic Mirror will now be described in more detail.

Table 7.3	Methods Available for the Monitoring of Beam Characteristics							
Instrument	**Characteristic**							
	Ref.	Power	Diam.	Mode	Wander	Dirt**	Refl.	Response Time
LBA	2	✓	✓	✓	✓	✓	✓	Fast
Perforated Mirror	3	✓	✓	✓				Fast
Chopper Devices	4,5	✓	✓	✓				Fast
Heating Mirrors	6	✓					✓	Slow
	7	✓	✓	✓	✓			Slow
	8	✓	✓	✓	✓			Fast
Heating Wire	9	✓	✓	✓				Slow
Photon Drag in Ge*	10	✓	✓	✓				Fast
Piezoelectric*	11	✓	✓	✓				Fast
Heating Gas	12	✓						Slow
	13	✓						Slow
	14	✓						Fast
Optical Scattering	15	✓				✓		Fast
Acoustic Signals	16	✓	✓	✓	✓	✓	✓	Fast
* In-process only if used with a beam splitter. **Mirror or lens fouling								

7.2.1.1. The Laser Beam Analyser (LBA) (2):

The laser beam analyser consists of a reflecting molybdenum rod which is rotated fast through the beam. The reflections off the rod are measured by two pyroelectric detectors placed as shown in Fig. 7.7.a. The two detectors pick up signals proportional to the power on two simultaneous orthogonal passes of the beam, as illustrated. It is this ability of the instrument to collect the power distribution within around 1/100th of a second in two dimensions simultaneously with only 0.1% beam interference that has made it one of the more popular beam measuring instruments in a laser facility. It is often fitted after the output window and before the shutter assembly so that the beam can be monitored even when it is not being used. The

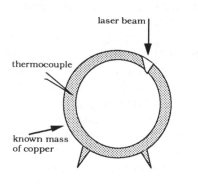

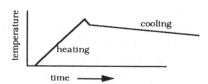

Fig. 7.6. Diagram of an absolute blackbody calorimeter

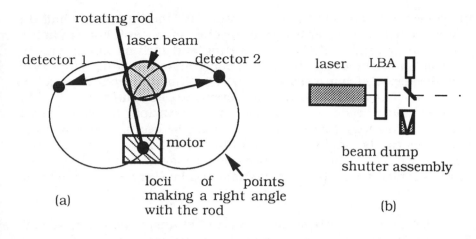

rotating rod

laser beam

detector 1   detector 2

motor

locii   of   points
making a right angle
with the rod

(a)

laser   LBA

beam dump
shutter assembly

(b)

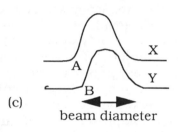

X

A

Y

B

beam diameter

(c)

Fig. 7.7. The Laser Beam Analyser (LBA). a)
The principle whereby two orthogonal passes
can be made simultaneously. b) An arrangement for continuous viewing of the beam. c)
Example of the oscilloscope output.

arrangement is illustrated in Fig. 7.7.b. The signals from the instrument
can be displayed on an oscilloscope or passed to a computer for further
analysis. The data which can be gained from the signals, which are
illustrated in Fig. 7.7.c, are:

1. Overall power: measured from the integral under the curve or
    RMS value.
2. Beam diameter: measured from the $1/e^2$ position of the power
    rise.
3. Beam wander: measured from any variation in the relative rise
    positions A and B of the two traces.
4. Mode structure: measured from shape of curves, in particular
    comparison with previous mode structures can be made to
    check on cavity tuning, or fouling of cavity mirrors.
5. Any instantaneous variations in the above properties. It is quite
    surprising how much power vibration and beam dilation there is
    in the beam from some lasers.

The instrument can be used to inspect the focus of the beam but that,
of course, can not be done in-process. Detailed analysis of the signals
from an LBA is given in reference (17).

Instruments which only measure power have measured only half the story. The beam diameter is equally or more important. For example it should be remembered that the cutting and melting ability can be correlated with the group (P/VD); the penetration in welding with P/VD² and the depth of hardness with P/√VD. So the versatility of this instrument and the small beam interference are its main strengths. The rotating hollow needle is a variation on this principle. It scans a small hole (approximately 23μm diameter) across the beam. Power distribution maps can be produced by a multipass scanning system. A map takes about 5s to achieve. This is more of a diagnostic tool than an in-process control tool. It is able to map the focal power distribution at low powers.

7.2.1.2. Acoustic Mirror Beam Monitor (16): This instrument may serve to illustrate the surprises in store for laser engineers. It was found by Weerasinghe and Steen in 1984 that high frequency acoustic signals were generated in mirrors which were reflecting laser radiation. The arrangement is shown in Fig. 7.8. A typical signal is shown in Figs. 7.9 and 7.10. Perhaps the strangest part was the long term oscillation in the RMS signal shown in Fig. 7.11. The signal responds to the power, beam diameter, position on the mirror, state of tuning of the laser and even the gas composition in the laser cavity! What is considered to be happening is that radiation falls on the mirror and in the action of reflecting the power, some power is absorbed which instantaneously heats the surface atoms. This causes an expansion and hence a stress wave which passes through the mirror (and water cooling at the back if necessary) to the piezo electric detector. The concept of the stress being caused by photon pressure is not considered likely since, as seen in Chapter 2, the stress would be relatively very low and would be expected to increase with the reflectivity of the mirror rather than what happens which is the opposite. Thus the instrument is recording only the variation in power - not the absolute power. With this picture in mind the phenomena observed with this instrument can be understood. Firstly an increase in laser power is usually associated with an increase in the power variation. Some lasers

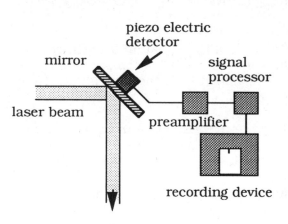

Fig. 7.8. The acoustic mirror arrangement.

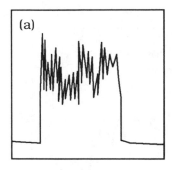
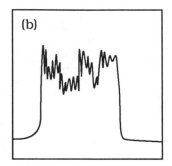

Fig. 7.9. Diagrams illustrating the form of the raw signal from the acoustic mirror (a) and the form of the signal after passing through a low pass filter (b).

are more stable than others. For example a slow flow laser will give a smaller signal than a fast axial flow laser for the same power. A beam of larger diameter will have a higher surface thermal stress than one of smaller diameter and hence the signal would rise if the diameter increases with the same power fluctuations. So the question becomes why should there be a fluctuation in the power of a laser beam?   Fig. 7.10 shows that the signal is made of two frequency components, one at around 100kHz and one around 1MHz. The low frequency component varies with the gas mixture and state of tuning of the cavity. In fact for certain gas mixtures it will disappear altogether. This low frequency element correlates with the frequency observed for the brightness of the plasma in the laser cavity and is considered to be due to the plasma attachment frequency at the cathodes. The higher frequency component is thought to be a function of the cavity design and corresponds to the photon oscillations in the cavity. The time to exhaust the inverted population and the time to rebuild it results in the laser virtually spitting power rather than being truly continuous. Fig. 7.11 shows that the RMS signal with a 10ms time constant has a slowly oscillating intensity not detected by the flowing cone calorimeter

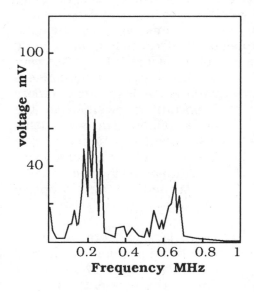

Fig. 7.10. Frequency spectrum of the raw signal from the acoustic mirror showing two peaks one in the 100kHz region and one towards the MHz region (18).

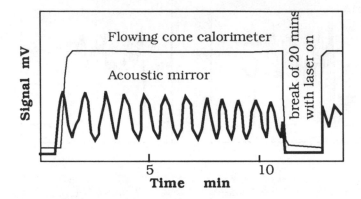

with a 10s time constant. Since the oscillations in signal strength take over a minute this should have been detected by both instruments if it had been due to a power variation. The LBA also showed no power variation but by very careful observation of the LBA signal a small oscillation in beam diameter was observed. The acoustic mirror picks up this beam dilation - or rather power variation - strongly. It is believed, but not yet proven, that as the laser warms up, so the cavity expands and the optical oscillations in the cavity form standing waves on the optic axis when the cavity is an exact number of wavelengths long and slightly off axis when it is not an exact number of wavelengths long. This very small variation would cause a variation in beam size but more important a variation in the beam stability. It does mean this instrument, with no beam interference, could be a tool to identify automatically when a laser is warmed up and properly tuned.

There is a variation in signal with beam position on the mirror as shown in Fig. 7.12. This is due to the different distances the signal must travel within the mirror to get to the detector, and also possibly due to some of the beam missing the mirror at the edges.

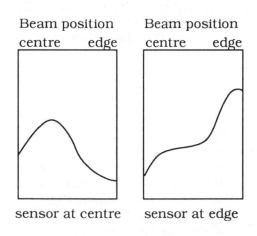

Fig. 7.12. Variation of RMS signal from the acoustic mirror with beam position on the mirror.

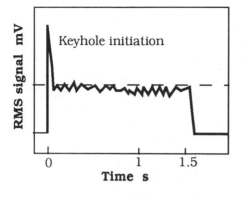

Fig. 7.13. Acoutic signal obtained from a good weld with steady keyhole.

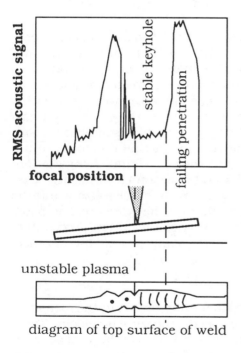

diagram of top surface of weld

Fig. 7.14. Variation of signal obtained by welding a sloped sample. very high signals are obtained when the keyhole fails and there is a large back reflection.

Thus the acoustic mirror is capable of measuring instantaneously and without any beam interference variations in the following:
1. Power.
2. Beam diameter.
3. Beam position on mirror.
4. State of laser tuning.
5. Approximate gas composition in cavity.
6. Back reflections from the work piece.

This formidable collection of data is unfortunately too much of a cornucopia of data. It is difficult to unravel one variable from another when the only analytic variables are signal strength and frequency. It seems to be a case waiting for high speed pattern recognition which may possibly come with neural logic.

The last item on the above list was the detection of back reflection signals. If the acoustic mirror is mounted near the workpiece it will receive back reflections from the workpiece, which should be diagnostic of the process condition. These back reflections are strongly fluctuating signals and on the whole they cover the whole mirror surface, so their thermal stress variations are very large compared to the relatively stable, but far more powerful, laser beam. A 5% variation in a 2kW beam is 100W while a back reflection of 200W may have a 100% variation. Since the reflectivity of many materials is over 60% it can be appreciated that this calculation is modest and that the acoustic sensor would be expected

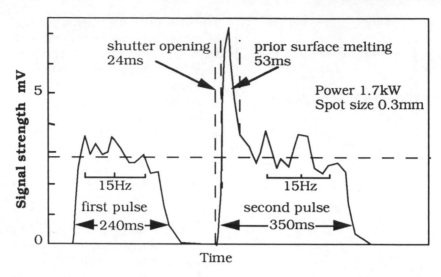

Fig. 7.15. The acoustic signal obtained from two laser pulses on the same spot.

to measure back reflections more strongly than the main beam. Some of the results are illustrated in Figs. 7.13 and 7.14. In Figs. 7.13 and 7.14 the use of the acoustic mirror as a keyhole sensor is illustrated. In Fig. 7.15 the signals from a stationary spot being melted by a laser are shown. The first pulse on a flat plate has a fairly uniform signal with a possible 15Hz variation in it. The second pulse on the same spot has a high initial peak due to the strong reflection from the resolidified smooth concave cavity left from the first pulse. It appears that this took around 53ms to remelt and then the signal again showed a faint 15Hz variation. Calculations by Postacioglu, Kapadia and Dowden (19) indicated that the expected natural wave frequency on pools of this size would be around 15Hz. Are we looking at (listening to) the waves on the pool? Certainly the acoustic mirror would be a very significant in-process tool if such intimate data could be obtained without any process interference.

7.2.2. Monitoring Work Table Characteristics

The work table variables are fairly straight forward as in the measurement and control of position, traverse speed or nozzle gas velocity. Table 7.1 listed many of the well known techniques. However the focal position is crucial and more subtle to measure. It will now be discussed in more detail.

7.2.2.1. Monitoring Focal Position: The focal position needs to be measured in real time because the workpiece may warp slightly during processing or the part may be contoured and the programming of the line to be followed may be simplified if the height above the workpiece

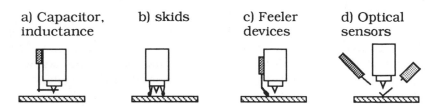

Fig. 7.16. Examples of various ways of sensing the focal position.

is controlled in-process rather than left to precise and time consuming programming. There are several signals which are used currently for sensing and controlling the height of the nozzle above the workpiece. They are shown in Fig. 7.16.

a) Inductance(20) and capacitance(21) devices can be made to look very neat and built as part of the nozzle. Their problems arise from being only applicable to processing metal and to suffering signal changes near edges and due to water or debris on the surface, particularly oxides. The edge problem may be acute in cutting nested articles. Plasma in welding may also affect the signals.

b) Skid devices have the nozzle riding on the workpiece. To avoid bouncing and recoil due to the air cushion effects from the nozzle when cutting without full penetration, the nozzle is loaded with a kilogram or so of force. This means that cutting soft material may cause a scar on the surface. Also cutting small parts is difficult by this method. The nozzle is usually made with a slotted piece of tungsten carbide or a set of roller balls. This system does have the advantage of allowing the nozzle to be very close to the cut surface and hence gives good aerodynamic coupling between the jet and the slot.

c) Feeler devices are sensitive lightly loaded lever systems which operate a piston in an inductive pot. They are fast acting and, with properly shaped feeler skids, can be built to work close to edges. The advantages of this and the skid sensing systems are that they are not sensitive to the material being cut nor water on the surface.

d) Optical height sensors have been built based on a He/Ne or diode laser beam illuminating a spot very near the interaction point. The image of this spot is focussed by a telescope onto an array of detectors capable of sensing any variation in image location. The probe beam can be modulated and the detector fitted with a filter to only allow that frequency to be measured. By this means the signal can be separated from the light emitted by the process. An alternative to this elaborate system was one used by Li (22) using an IR transceiver. The intensity of the reflected signal varies strongly with height. However this small electronic component has to be mounted further from the interaction zone than the probe beam system. Both optical devices have the advantage that they are suitable for all materials, they are not sensitive to the presence of edges and are relatively insensitive to surface water or debris.

7.2.2.2 Seam following: In butt welding there is a need to follow the joint line. In laser welding, with the narrow fusion zone associated with laser welds, there is a need for a seam following system which is accurate and fast. Several systems have been suggested. There is the optical system of Lucas (23). In this a CCD camera, using suitable filters, is able to pick up the laser line which is formed by passing a He/Ne or diode beam through cylindrical optics. The straightness of the line is analysed by a computer and the location of the seam found within a few μs. The control system has been tested and proved to work in TIG welding (24). An alternative is to scan the beam, as with Oomen's method (25). Inductive sensors have also been developed which look at the variations in magnetic field around a joint (20).

7.2.3. Monitoring Workpiece Characteristics

7.2.3.1. Temperature: The temperature of the workpiece is important in determining the extent of a transformation hardening process or in the level of dilution expected during laser cladding. There are several methods for examining temperature. There is a straightforward pyrometer looking at the interaction zone, as in Bergmann's method (26) for controlling the transformation hardening process. There are CCD cameras which can look at the welding process and with appropriate software compute the size of the weld pool from which the penetration might be determinable (27). Infrared red scanning pyrometers have been used to measure the thermal profile around the event.

Optical intensity has been viewed through the mirror by Zheng (28) using a fibre mounted in the last beam guidance mirror. His signals indicated the extent of dross adhesion and the formation of striations during laser cutting. Olsen (29) using a beam splitter was also able to witness these detailed events during cutting. Miyamoto et al (30) had a system for viewing the cut face during cutting. Their equipment was used to take some very informative film and could be adapted to be less intrusive using fibres.

7.2.3.2. Keyhole Monitoring: The monitoring of the keyhole by an acoustic mirror has been discussed in Section 7.2.1. It can also be achieved using the "see through" mirror of Zheng et al (28) . There is also an electric signal, identified by Li et al (31) (the Plasma Charge Sensor) which can diagnose the general health of a laser weld while it is occurring. The plasma intensity is also an indicator and has been observed by many using optoelectric sensors, for example the work of Beyer (32) in Aachen. Others have simply listened to the welding process by microphone.

7.2.3.3. Dilution Monitoring: There are many new variables introduced during laser cladding. For example, there is the powder feed rate, the height of the clad track, the temperature of the substrate and the dilution of the clad. Li (33) invented a device for in-process monitoring the dilution of a clad layer based on a micro inductance sensor. The signal varied with the level of dilution in certain alloy systems, but it also varied with change of height and temperature. Thus for the signal to be useful there has to be a computation between several signals. This is a greater order of complexity than the other sensors just considered, but indicates the way the control of laser processing may probably progress.

7.2.3.4. Spark Discharge Monitoring: The angle at which the sparks leave a cut and the cone angle of the discharge is indicative of the health of the cutting process. ETCA (34) in Paris has developed a TV camera system to look at the spark discharge from the underside of the cut. The viewing angle is transverse to the cutting direction. The image is passed to an image processor from which control data is elicited and used to control the process. The process being a multi variable, multi option one means that some form of "intelligent" processing had to be developed by ETCA. The general arrangement is shown in Fig. 7.17. Its weakness is that the instrumentation has to be able to see beneath the workpiece, which is not always very convenient.

7.2.3.5. Clad Bond Condition Sensor: A photo electric sensor was used by Li et al (35) to look at the radiation from the melt pool during cladding. The signal was around 3 to 4 times higher than normal when the clad was not bonding correctly.

7.3. In-Process Control

It is one step to gain the diagnostic signals, the next is to use them in a closed loop control system. These control systems come in two basic varieties. There are those which are one to one. That is, for example, the power is measured and the power is controlled. It is obvious what to do. If the power is too low, raise the current to the discharge tubes. If it is too high, do the opposite. The alternative control systems occur where there are many diagnostic signals and many interrelated operating conditions to be adjusted. This latter type of control system requires decision making software which could constitute "intelligent" process- ing as discussed later in Section 7.4. For the straight control loops of

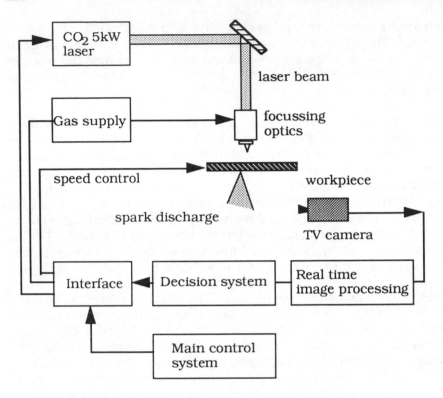

Fig. 7.17. Example of the ETCA smart laser cutting system (34).

the first type which do not require a decision as to what has to be altered but only by how much, we will discuss the in-process control of power and temperature, by way of example.

7.3.1. In-Process Power Control

Li et al (35) were able to monitor the beam power using the LBA. They showed that the RMS signal with a 10ms time constant was linearly proportional to the observed power as detected by a flowing cone calorimeter with a 10s time constant. This signal was passed into a PID controller (Proportional, Integral and Differential) and the PID constants adjusted to give fast damping. The PID generated control signal was used to control the potential to the cavity tubes and thus the current and so the laser power. The results are shown in Fig.7.18. The warm up time of the 2 kW CO_2 Control laser used, was around 12 minutes during which time the power varied sufficiently that the laser could not be used reliably. For the same warm up period, but using the control system, the power was attained within a minute and was then stable to within ±5W for 1.6kW. What was surprising in these results was that the mode

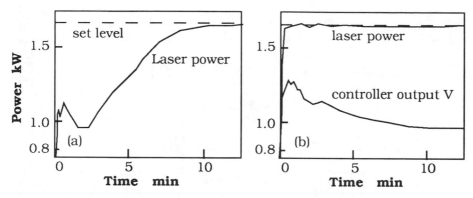

Fig. 7.18. The effect of in-process control of the laser power. a) The uncontrolled power during the warm up of the Control Laser 2kW machine. b) The same process but using feedback control of the power.

structure during this period was also stabilised. Working with ultra stable beams is something new to material processing engineers. The weakness of this system is that the controlled power must be below the maximum power so that the control can call on higher currents without causing an overload situation.

7.3.2. In-Process Temperature Control

In transformation hardening and laser cladding it is important to know or control the temperature of the substrate since this determines the extent of hardening or the level of dilution. Bergmann (26) using a pyrometer to measure the temperature at the heated zone was able, via a control circuit to control the speed of the table and so control the temperature of the workpiece. The depth of hardness achieved was uniform. This is a simple and reliable control system for a real industrial problem. It could have problems when there is a burning of the graphite or other coating used to aid absorption, which one might expect to interfere with the signal. This was overcome by using an adaptive temperature control algorithm (36).

7.4. "Intelligent" In-Process Control

If one wishes to control a whole process such as welding, cutting or cladding, then the control system will have to handle many different in-process sensing signals and will be faced with several control options. For example in cladding there will be signals from the height, tempera-ture, dilution, powder feed rate, laser power, traverse speed, height of clad and possibly others. The control options are to vary the power, powder feed, focal position and traverse speed. Thus the process will need straight control loops on power, powder feed, traverse speed etc in

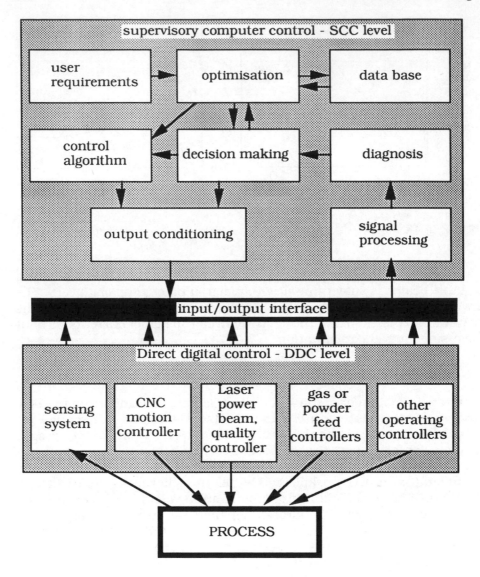

Fig. 7.19. Diagram of the logical framework for an "intelligent" control system (38).

order to have control over these variables and to know they will not wander; but on top of this an overall diagnostic package is required to determine what is wrong, if anything, and what to alter. This decision is then passed to another package to determine by how much that parameter should be altered. The "intelligent" part is for this decision making process to have its own feedback so that it is capable of making faster and more accurate diagnosis and adjustments. The logical frame for all this is illustrated in Fig. 7.19. The wider view has been seen in the arrangement of the ETCA cutting system (Fig. 7.17).

Table 7.4	Examples of Signals, Diagnosis and Responses for Probability Matrix				
	Sensed Signal		Possible Diagnosis		Possible Responses
s1	Clad bead too rough	d1.1	Powder flow rate too high	r1	Decrease powder flow rate
s2	Clad bead discontinuous	d1.2	Powder particle velocity too low	r2	Increase powder delivery gas pressure
s3	Bad laser mode structure	d1.3	Bad mode structure	r3	Decrease nozzle gas pressure
s4	Nozzle temperature too high	d1.4	Lens cracked	r4	Nozzle gas pressure too high: set alarm
s5	Clad bead too narrow	d1.5	Nozzle gas pressure too high	r5	Powder delivery pressure too high: set alarm
s6	Clad is too thin	d1.6	Traverse speed too high	r6	Laser power too high
		d1.7	Power too high	r7	Traverse speed too high
		d2.1*	Powder delivery gas pressure can not be altered on line		
		d2.2*	Nozzle gas pressure can not be altered on line		
*Some control restriction affecting the diagnosis					

The diagnostics could be arranged as in Li's method (37) as a probability matrix for the solution of fuzzy logic problems. An alternative is the use of neural logic. As an example of a fuzzy logic matrix some of the diagnostic options for the sensor signals and responses are shown in Table 7.4.

These options will be seen by the computer as a probability matrix illustrated in Table 7.5. The sensing logic will have indicated a "true" or "false" situation regarding the state of each symptom. This matrix example, which is only a partial matrix for the full problem, indicates that the most probable diagnoses are d1.1, d1.2 or d1.5.

The diagnoses are selected when their probability rating exceed a threshold amount, say 0.5. They are then fed into a decision probability matrix illustrated in Table 7.6. Thus the response should most probably be r1, or failing that r5 or failing that r4. It may be that r1 can not be altered any more in a given direction.

The probability values in the probability matrices are variables which could be altered by the decision making program. Thus if r1 is altered and it is found not to work then the probability of 1 would be reduced by a calculated amount. In this way the program would "learn" a better probability table and therefore a swifter and more accurate response. The subject is developed in greater depth by Li (22).

Table 7.5		Diagnosis Probability matrix					
Signal	s1	s2	s3	s4	s5	s6	sum d1.x
State	true	true	false	false	false	true	
d1.1	**0.5**	**0.5**	**0**	**0**	**0**	**0**	**1.0**
d1.2	**0.4**	**0.5**	**0**	**0**	**0**	**0**	**0.9**
d1.3	0.2	0.1	0.3	1	0	0	0.3
d1.4	0.2	0.2	1	0.3	0	0	0.4
d1.5	**0.3**	**0.4**	**0**	**0**	**0**	**0**	**0.7**
d1.6	0	0.3	0	0	0.5	0	0.3

Table 7.6		Response Probability Matrix				
Diagnosis	d1.1	d1.2	d1.5	dn.1	dn.2	
Probability	(1)	(0.9)	(0.7)	(1)	(1)	sumrx
State	true	true	true	true	true	
r1	1	0	0	0	0	1.0
r2	0	0.5x0.9	0	-1	0	-0.55
r3	0	0	0.5x0.7	0	-1	-0.65
r4	0	0	0.5x0.7	0	0.5	0.85
r5	0	0.5x0.9	0	0.5	0	0.95

That is only half the problem. The next half is to decide by how much r1 should be adjusted. In fact by how much should the powder flow rate be decreased. This is standard control theory which is designed to make these alterations without introducing instability into the control system. A rule based adaptive control strategy is given in (22).

7.5. Conclusions

The laser is an ideal partner for automation due to the ease for gaining in-process signals regarding the state of the process. The gains to be made by having a fully automatic self correcting and improving process are currently only partly perceived. For factories to work through the night with the lights out and the heating off; for the factory owner to play golf all day and the machines to work all day and night and for the product quality to be uniformly high may seem a utopia. The laser is becoming well suited to help bring this about. Then gentle reader what will you do? There is the challenge of positive use of leisure - one could always start a laser job shop or something!

References

1. Craig.J.J. "Adaptive control of mechanical manipulators" publ Adison Wesley 1988 Mass USA.
2. Lim.G.C., Steen.W.M. "The Measurement of the Temporal and Spatial Power Distribution of a High Powered CO_2 Laser Beam" Optics and Laser Technology June 1982 pp149-153.
3. Spalding.I.J. "High Power Laser Beam Diagnostics pt.I" Proc 6th Int Gas Flow and Chemical Laser Conf. (GLC6), Jerusalem, Israel Sept 1986 publ Springer-Verlag pp314-322.
4. Sepold.G., Juptner.G., Rothe.R. "Remarks on Deep Penetration Cutting with a CO2 Laser" Paper A-29 Proc Intl Conf. on Weld. Res. 1980 Osaka, Japan JWRI publ.
5. Oakley.P.J. "Measurement of Laser Beam Parameters" IIW DOC IV-350-83 1983.
6. Rasmussen.A.L. "Double Plate Calorimeter for Measuring the Reflectivity of the Plates and Energy in Beam of Radiation" US Patent 3,622,245 Dec 1971.
7. Mansell.D.N. "Laser Beam Scanning Device" US Patent 3,738,168 June 12 1973.
8. Davis.J.M. Peter.P.H. "Calorimeter with a Highly Reflective Surfacefor measuring Intense Thermal Radiation" Appl. Opt. Vol 10 No.8 Aug 1971.
9. Toshharu Shirakura et al "Methods and Apparatus for Measuring Laser Beam" US Patent 4,474,468 Oct 2 1984.
10. Gibson.A.F., Kimitt.M.F., Walker.A.C. Appl. Phys. Lett. 17 1970.
11. Satheesshkumar.M.K., Vallabhan.C.P.G. "Use of a Phot-Acoustic Cell as a Scientific Laser Power Meter" J. Phys E Sci Instr. Vol 18 1985 pp435-436.
12. Ulrich.P.B. "Power Meter for high Energy Lasers" US Patent 4,381,148 April 26 1983.
13. Miller.T.G. "Power Measuring Device for Pulsed Lasers" US Patent 4,325,252, April 20 1982.
14. Shifrin.G.A. "Absorption Radiometer" US Patent 3,487,685 Jan 6 1970.
15. Crow.T.G. "Laser Energy Monitor" US Patent 4,424,581 Dec 30 1980.
16. Steen.W.M., Weerasinghe.V.M. "Monitoring of Laser Material Processing" Proc SPIE conf. paper 650 22 Innsbruck April 1986 publ by SPIE PO Box 10 Bellingham, Washington USA proc vol 650 pp160-166.
17. Lim.G.C., Steen.W.M. "Instrument for the Instantaneous in situ Analysis of the Mode Structure of a High Power Laser Beam" J. Phys.E. Sci.Instr. (1984) vol 17 pp999-1007.

18. Willmott.N.F.F., Hibberd.R., Steen.W.M., "Keyhole/Plasma Sens
    ing System for Laser-Welding Control System" Proc Int Conf on
    Applications of Lasers and Electro Optics ICALEO '88 Santa
    Clara Calif USA Oct/Nov 1988 publ Laser Material Processing
    Springer-Verlag IFS publ. with LIA pp109-118.
19. Postacioglu.N., Kapadia.P, Dowden.J., "Capillary Waves on the
    Weld Pool in Production Welding with a Laser" Journ. Phys. D.
    Applied Phys. Vol 22 pp1050-1061 1988.
20. Goldberg.F. "Inductive Seam Tracking Improves Mechanisation
    and Robotic Welding" Proc Automation and Robotisation of
    Welding, Strasbourg France (1985).
21. Hanicke.L."Laser technology within the Volvo Car Corporation"
    Proc 4th Int Conf. Lasers in Manufacturing (LIM4) B'ham UK
    May 1987 publ IFS(publ) Bedford, UK and Springer-Verlag,
    Berlin FRG 1987 pp49-58.
22. Li.L Ph.D. Thesis, London University 1989.
23. Lucas.J., Smith.J.S., "Seam Following for Automatic Welding"
    Proc SPIE Conf. Laser Technologies in Industry vol 952 ed.
    O.D.D.Soares June 1988 Porto, Portugal pp559-564.
24. Sloan.K., Lucas.J. "Microprocessor Control of TIG Welding Sys-
    tems" IEE Proc. pt.D 1,pp1-8 1982.
25. Oomen.G., Verbeek.W. "A Real Time Optical Profile Sensor for
    Robot Arc Welding" Proc Intelligent Robots ROVISEC 3 Cam-
    bridge Mass (1984).
26. Rubruck.V., Geisler.E., Bergmann.H.W., "Case Depth Control for
    Laser Treated Materials" Proc 3rd Europ. Conf. Laser Treatment
    of Materials ECLAT'90 Erlangen, Germany , publ Sprechsaal,
    Coburg, Germany, Sept 1990 pp207-216.
27. Juvin.D., de Prunelle.D., Lerat.B. "SAO par Imagerie: Un
    Algorythme de Vision du Bain de Soudure pour le Controle et la
    Conduite du Procede Soudage TIG" Proc Aut des procedes de
    Soudage, Grenoble, France (1986).
28. Zheng .H.Y., Brookfield.D.J., Steen.W.M. "The Use of Fibre
    Optics for In-process Monitoring of Laser Cutting" ICALEO'89
    Orlando, Florida, USA 12-22 Oct 1989 140/LIA vol 69 pp140-
    154.
29. Olsen.F. "Investigations in Methods for Adaptive Control of Laser
    Processing" Opto Electronik Magazin 4,2.
30. Miyamoto.I, Ohie.T., Maruo.H. "Fundamental Study of In-Process
    Monitoring in Laser Cutting" Proc CISFFEL 4 Cannes, France
    (1988).
31. Li.L , Qi.N, Steen.W.M., Brookfield.D.J. "On Line Laser Weld
    Monitoring for Quality Control" Proc ICALEO'90 conf Nov 1990
    Boston, USA publ. LIA Tulsa, Oklahoma USA. 1991 pp411-421.

32. Beyer.E."Plasma Fluctuation in Laser Welding with CW CO_2 Laser" Proc ICALEO'87 San Diego USA May 1987 publ IFS publ and Springer-Verlag in assoc LIA Toledo USA 1988 pp17-23.

33. Li.L., Steen.W.M., Hibberd.R.D., Brookfield.D.J. "In-process Monitoring of Clad Quality using Optical Method" Proc SPIE conf, Hague March 1990 vol 1279, pp89-100.

34. Burg.B. "Smart Laser Cutter" Proc SPIE conf Innsbruck, Austria April 1986 vol 650 ed. Schoucker publ. SPIE, Bellingham, Washington USA 1986 pp271-278.

35. Li.L. , Hibberd.R.D., Steen.W.M. "In-Process Laser Power Moni toring and Feedback Control" Proc 4th Int Conf on Lasers in Manufacturing LIM4 ed W.M.Steen Birmingham UK May 1987 publ IFS publ Ltd. Bedford UK pp165-175 1987.

36. Drenker.A., Beyer.E., Boggering.L., Kramer.R., Wissenbach.K. "Adaptive Temperature Control in Transformation Hardening" Proc 3rd Europ. Conf on Laser Treatment of materials ECLAT'90 Erlangen, Germany Sept. 1990 publ Sprechsaal, Coburg, Germany, pp283-290.

37. Li.L., Steen.W.M., Hibberd.R.D., Weerasinghe.V.M. "Real Time Expert System for Supervisory Control of Laser Cladding" Proc ICALEO'87 San Diego USA May 1987 publ IFS publ and Springer-Verlag in assoc LIA Toledo USA 1988 p9-16.

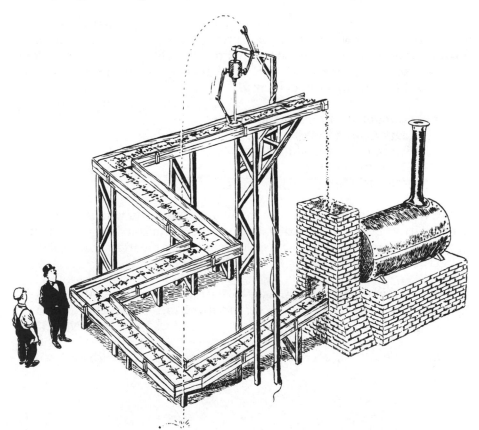

Chapter 8

Laser Safety

" He saw; but blasted with excess of light, closed his eyes in endless night"
Thomas Gray 1716-1771 The Progress of Poesy 1757, iii.2 (Milton)

8.1. The Dangers

All energy is dangerous, even gaining potential energy walking up stairs is dangerous! The laser is no exception, but it poses an unfamiliar hazard in the form of an optical beam. Fortunately, to date, the accident record for lasers is very good, but there have been accidents. The risk is reduced if the danger is perceived.

The main dangers from a laser are:
1. Damage to the eye.
2. Damage to the skin.
3. Electrical hazards.
4. Hazards from fume.

These risks can be made safe by following standards which have been laid down by various authorities. The principle standards at present are:
Safety of laser products and equipment classification and user guides - general laser safety -- BSEN 60825-1
IEC 825-1
Safety of machines using laser radiation to process materials - safety of laser material processing machines-- ISO DIS 11553
CENprEN 31553

" In the multitude of counsellors there is safety" Proverbs 11 v14.

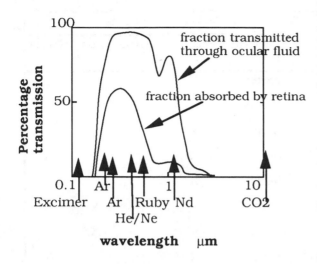

Fig. 8.1. Spectral transmissivity of the ocular fluid and the absorptivity of the retina.

These standards give guidance and rules concerning: engineering controls, advice on personal protective equipment, administrative and procedural controls and special controls. Class 4 laser installations, which is nearly all material processing systems, should also have a laser safety officer (LSO) who should see that these guidelines are observed. Following these rules the laser installation will be safe. Breaking the rules may result in an accident.

8.2. The Safety Limits

8.2.1. Damage to the Eye

The ocular fluid has its own spectral transmissivity as shown in Fig. 8.1. It indicates that there are two types of problem with radiation falling on the eye. There is potential damage to the retina at the back of the eye and potential damage to the cornea at the front of the eye. Radiation which falls on the retina will be focussed by the eye's lens to give an amplification of the power density by a factor of around 10^5. This means that lasers with wavelengths in the visible or near visible wave band (Ar, He/Ne, Nd-YAG, Nd-glass) are far more dangerous than those outside that band (CO_2, Excimer). The nature of the threat from different lasers is listed in Table 8.1.

Safe exposure limits have been found by experiment and they are listed as the Maximum Permissible Exposure levels (MPE levels). These levels are plotted in Figs. 8.2 and 8.3 for retinal and corneal damage. At power density and times greater than these safe limits damage may occur due to boiling or at higher levels to explosive evaporation. The boiling limit is the reason for the very low levels of power which the eye can tolerate. For example a 1mW He/Ne laser with a 3mm diameter beam would have a power density in the beam of $(0.001 \times 4)/(3.14 \times 0.3 \times 0.3) = 0.014W/cm^2$. On the retina this would be 0.014×10^5 W/cm^2. A blink reflex at

Table 8.1		Basic Laser Biological Hazards				
Laser Type	Wavelength μm	Biological Effects	Skin	Cornea	Lens	Retina
CO2	10.6	Thermal	X	X		
H2F2	2.7	Thermal	X	X		
Erbium-YAG	1.54	Thermal	X	X		
Nd-YAG	1.33	Thermal	X	X	X	X
Nd-YAG	1.06	Thermal	X			X
GaAs Diode	0.78-0.84	Thermal	**		X	
He/Ne	0.633	Thermal	**		X	
Argon	0.488-0.514	Thermal photochem	X			X
Excimer: XeF	0.351	Photochem	X	X	X	
XeCl	0.308	Photochem	X	X		
KrF	0.254	Photochem	X	X		

** Insufficient power

this level would only allow a 0.25s exposure, which is the MPE level for a class 2 laser. Notice that the calculation assumes that all the radiation can enter the pupil of the eye. Thus it is common practice to ensure that working areas around lasers are painted with light colours and are brightly illuminated - not so with holographic laboratories and others involved with photography, of course!

The hazard zone around a laser is that in which radiant intensities exceed the MPE level. These zones are known as the Nominal Hazard Zone (1). The size of the zone can be calculated based upon the beam expansion from the cavity, or lens, or fibre, or from diffuse or specular reflection from a workpiece. For example, consider a 2kW CO_2 laser beam with a 1 mrad divergence. The MPE level for safe direct continuous viewing of the beam (not that much would be seen with IR radiation!) is when the level falls to $0.01W/cm^2$. This would occur when the beam has expanded to 504cm diameter - a distance of 5020 m away, around 3 miles! This means that precautions must be taken to avoid the beam escaping from the area of the laser by installing proper beam stops, screens for exits and enclosed

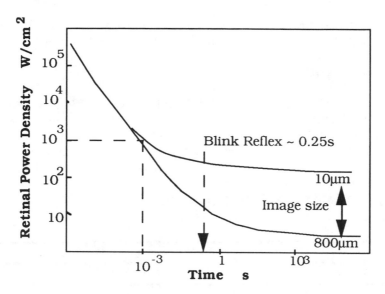

Fig. 8.2. Approximate exposure limits for the retina

beam paths. Similar calculations for a 500W CO_2 laser give a Nominal Hazard Zone for diffuse reflections of 0.4m. Therefore it is necessary to wear goggles when near a working laser. As a general rule, **"Never look at a laser beam directly"** - it is like looking down a gun barrel.

8.2.2. Damage to the Skin

There are also MPE levels for skin damage. These are far less severe than for the eye and so are essentially irrelevant. The laser is capable of penetrating the body at speeds as fast as that for steel and so the focussed beam needs to be seriously respected. Without meaning to trivialise the problem with skin effects, the damage done is usually blistering or cutting, neither pleasant but the wound is clean and will heal - unlike some eye damage. Incidentally a vein or artery cut by laser will bleed even though it is cauterised!

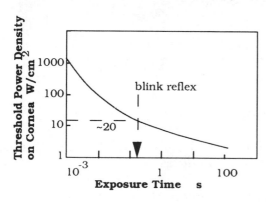

Fig. 8.3. Approximate damage threshold power densities for the cornea of the eye.

As a general rule: **"Never put parts of your body in the**

path of a laser beam". If an adjustment has to be made to the beam path do it by holding the edges of mirrors etc.

8.3. Laser Classification

Lasers are classified in BSI 4803 1983 and IEC 825 1984 according to their relative hazard. All lasers of interest to material processing will be classified as class 4 except some which are totally built into a machine in which there is no human access possible without the machine being switched off; so the list is somewhat academic for this book. Table 8.2. is a summary of the classification:

8.4. Typical Class 4 Safety Arrangements

Table 8.2.	Classification of Lasers
Class	Definition
1	Intrinsically safe $< 0.2\mu J$ in 1ns pulse or $< 0.7mJ$ in a 1s pulse
2	Eye protection achieved by blink reflex (0.25s) $< 1mW$ CW laser
3A	Protection by blink and beam size $<5mW$ with $25W/m2$ (e.g. an 16mm beam diameter from a 5mW laser)
3B	Possible to view diffuse reflection $< 2.4mJ$ for 1ns pulse or $< 0.5W$ CW visible
4	All lasers of higher power Unsafe to view directly, or by diffuse reflection May cause fire Standard safety precautions must be observed (Section 8.4.)

The following precautions are advised:

*   All beam paths must be terminated with material capable of withstanding the beam for several minutes.

*   Stray specular reflections must be contained.

*   All personnel in Nominal Hazard Zone must wear safety goggles. For CO_2 radiation they can be made of glass or perspex, in fact normal

spectacles may do, if the lenses are large enough.

*   Non involved personnel must have approval for entry.

*   There should be warning lights and hazard notices so that it is difficult (impossible) to enter the area without realising that it is being entered.

*   Extra care should be taken when aligning the beam.

*   There should be a Laser Safety Officer to check that these guide-lines are followed.

These guidelines are summarised in Fig. 8.4 of a typical laser material processing arrangement.

8.5.   Where Are the Risks in a Properly Set Up Facility?

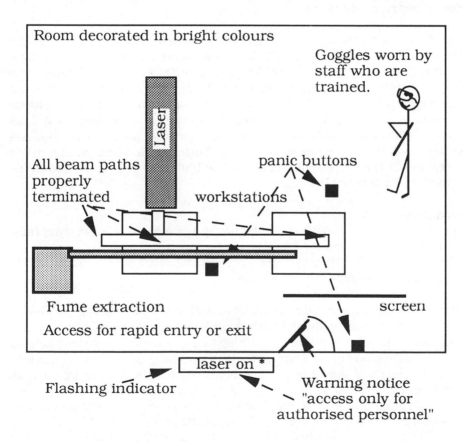

Fig.8. 4. Diagram illustrating safety features of a laser laboratory.

If the facility is properly designed then the beam is enclosed and all beam paths terminated so that the beam can not escape to do damage. A standard set up would have the beam focussed and pointing downward. As the beam expands after the lens the NHZ is considerably reduced. For robotic beam steering systems this NHZ is the distance to which the operators can approach the robot. This system is thus safe except in certain unlikely events. These events can be classified in a risk analysis tree. They would include breaking of the lens and total removal with loss of the nozzle, failure of a mirror mount and the mirror swinging free etc.. The essence of such an analysis is to devise a system for rapidly identifying an errant beam. This can be achieved by beam monitoring and/or enclosure monitoring. If the beam is monitored as leaving the laser but not arriving at the expected target then the system should immediately shut down. If a hot spot appears within the enclosure then again the system could shut down.

8.6. Electrical Hazards

Nearly all the serious or fatal accidents with lasers have been to do with the electric supply. A typical CO_2 laser may have a power supply capable of firing the tubes with 30,000 Volts with 400mA. This is a dangerous power supply and when working on it the standard procedures for electric supplies should be followed. The smoothing circuits contain large capacitors and so even when the power is switched off a fatal charge is still available and proper precautions to earth the system before working on it are essential. Panic buttons must be available at the laser and at the main exit. Access to the high tension circuit should be protected by interlocks.

As a general rule:
Do not enter the high voltage supplies without first carefully earthing the system.

8.7. Fume Hazards

The very high temperatures associated with laser processing are able to volatilise most materials and thus form a fine fume, some of which can be poisonous. With organic materials, in particular, the plasma acts as a sort of dice shaker and a wide variety of radical groups may reform into new chemicals. Some of these chemicals are highly dangerous such as the cyanides and some are potential carcinogens. It is necessary, as a general rule, to have a well ventilated area around the laser processing position as for standard welding. Some of the problems with cutting non metallic materials have been identified by Ball et al(2). These are shown

Table 8.3.	Main Decomposition Products from Laser Cut Non-Metallic Materials (2)				
Decomposition Products	Material				
	Polyester	Leather	PVC	Kevlar	Kevlar/Epoxy
Acetylene	0.3-0.9	4.0	0.1-0.2	0.5	1.0
Carbon monoxide	1.4-4.8	6.7	0.5-0.6	3.7	5.0
Hydrogen chloride			9.7-10.9		
Hydrogen cyanide				1.0	1.3
Benzene	3.0-7.2	2.2	1.0-1.5	4.8	1.8
Nitric dioxide				0.6	0.5
Phenyl acetylene	0.2-0.4			0.1	
Styrene	0.1-1.1	0.3	0.05	0.3	
Toluene	0.3-0.9	0.1	0.06	0.2	0.2

in Table 8.3. but it should be remembered that the volumes/s are not very large and represent a hazard only if much work is being done over an extended time.

As a general rule:
 " **The laser processing zone should be adequately ventilated".**

8.8. Conclusions

The laser is as safe as any other high energy tool and should be properly handled.   It is the responsibility of the user to learn how to handle it correctly.

References

1. Rockwell.R.J. "Fundamentals of Industrial Laser Safety" Industrial Laser Annual Handbook 1990 publ Penwell books, Tulsa, Oklahoma,USA pp131-148.
2. Ball et al. Industrial Laser Annual Handbook 1989 publ Penwell books, Tulsa, Oklahoma,USA p 23.

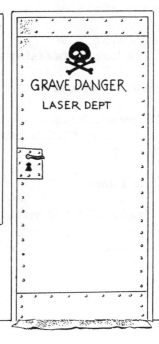

INTERNATIONAL SAFETY CODES
ON OPERATION OF LASERS.

1. GOGGLES MUST BE WORN AT ALL TIMES.
2. DO NOT USE THE LASER AS A
   CIGARETTE LIGHTER.
3. DO NOT DIRECT LASER BEAM AT
   CANS OF TINNED BEANS
4. IN THE EVENT OF SERIOUS INJURY
   TRY NOT TO MAKE A FUSS.
5. DO NOT FIDDLE ABOUT WITH
   THE LASER – IT'S EXPENSIVE!
6. ON NO ACCOUNT MUST THE
   LASER BE LENT TO POP GROUPS.
7. ALL DOGS TO BE KEPT ON LEADS
8. WHEN YOU'VE FINISHED USING
   THE LASER ROOM PLEASE LOCK
   UP AND REPLACE KEY UNDER
   THE MAT.
9. DO NOT FORGET TO BLOW
   OUT THE CANDLES.
10. DO NOT USE THE LASER AS A
    SUN LAMP.

GRAVE DANGER

LASER DEPT

Epilogue

"They are ill discoverers that think there is no land when they can see nothing but sea"
Francis Bacon (1561-1626) Advancement of Learning Book1 vii 5.

This book started with the extravagant claim that optical energy was a new form of energy and therefore should lead to a major advance in our quality of life, as has been the case with the mastery of other forms of energy. Thus the expectation was high. Hopefully the patient reader has seen in the chapters of this book that some of this expectation is beginning to shape up. It seems appropriate, therefore, to finish with a little thoughtful wander through the future possibilities for laser material processing, which it must be remembered is only one small part of the impact of optical energy.

The major developments are likely to be centred on the differences of optical energy with other forms of energy as well as any equipment developments which will alter the capital cost or processing capabilities. Some of the principle differences between optical energy and other forms of industrial energy are that current laser power densities are amongst the highest available to industry today, that optical energy is amongst the easiest forms of energy to direct and shape, that this power can be delivered with very little signal noise allowing a unique window into the process for automation and adaptive control, that the laser beam contains properties as yet unexplored by material processing engineers but which are contributing to the rapidly growing "optical" applications of lasers. These may spill over into material processing. Multiphoton events and interactions with radiation of very pure spectral frequency may allow chemical processing which is unexpected thermodynamically. Finally it is very likely that there is a wealth of applications so novel that we have not seen the possibilities yet - this was certainly the case with fire, steam, electricity and oil. Consider these points in turn.

Power Intensity

Quick and accurate control of the power intensity would allow precise thermal histories to be engineered. This would allow control of cooling rates, stirring action, time above certain temperatures and weld bead profile. New materials for surfacing applications can be expected based on increased solid solubility, in situ alloy formation and composites containing thermally sensitive material such as diamond, Si_3N_4 and others.

Power Transmission

Optical energy is one of the few forms of energy which can be transmitted intact through air or space. Currently there is a great thrust for fibre delivery systems for laser beams. This is throwing away one of the distinctive advantages of this form of energy. The ease of placing optical energy should lead to developments in time sharing of beams over considerable distances within factories. The placement of optical energy within existing tooling and machinery may lead to the saving of entire unit operations. The application of fibres will always be haunted by the limited power which can be delivered without destroying the focussability.

Power Shaping

No other form of energy can be shaped with such precision as optical energy. Holographic mirrors have been invented and developments are promising. Scanning systems are becoming more subtle. The future appears to lead to complex shaping of high powered beams allowing single shot processing of areas for surfacing, cladding, drilling or even cutting. The development of mode matching optics could lead to the conversion of poor modes to high quality Gaussian modes - even from fibres.

Automation

Laser power has so many ways of being monitored. The process is wide open to in-process sensing and the development to "intelligent" processing is only just out of reach. Surely this is where the largest future for the laser exists.

Beam Coherence

This is a property as yet unused by material processing engineers. Yet it has the property of measuring distance, even possibly penetration distance in-process. It has the property of allowing interference effects which would in turn allow unusually detailed surface heat sources such as fringe systems to be used. Surely someone must find a use for this at some time.

Beam Spectral Purity

This is another property hardly used in material processing except in isotope separation. A challenge or a red herring?

Multiphoton Events

This is something the laser has brought to the world which we have never contemplated before. The excimer with its ability to cut "cold" has shown the extraordinary activity of photons in quantity. There are many non linear events demonstrating that photons in bulk lead to unusual events; consider, Brouillon, Rayleigh and Raman scattering. Have we the possibility of an entirely new chemistry?

Frequency Related Events

Isotope separation, photodynamic therapy and laser enhanced plating all indicate a possible mine of new techniques awaiting the wit of the inventor.

Equipment Developments

If someone were to build a mass produced laser for half the current price, then firstly he would still make a large profit per laser; but secondly a whole new range of applications would become cost effective. In fact the laser market is an elastic market; one which would grow faster than the reduced profit from each laser, if prices were reduced - but it requires mass production, not the current hand built systems. Thus it seems the market is waiting for a wealthy laser manufacturer with a good design, clear vision and a cool nerve to make the substantial investment a mass production line requires.

High powered Nd-YAG systems pumped by diodes instead of flash lamps would possibly increase the power without increasing the cooling problems of the current lasers. Phase matched diode arrays appear possible as a high powered source. For example, if 500 of the present 10W diodes were to be linked so that they emit in phase, this would produce a single phase wide beam which should be focussable. It would be a 5kW plate laser; a machine which could be small, light, powerful and of adaptable shape - and therefore readily inserted into industrial machinery. Such a device is now on the market - a fist size laser!

The development of optics so that poor modes or even Gaussian modes could be focussed to near theoretical spots has sufficient market advantages that someone must be working on it.

Automatic beam guidance systems, which are currently receiving some research interest, will lead to the simple control of laser beams over long distances allowing a new field of applications to develop.

Unthought of Concepts

Analogy with the introduction of all other forms of energy leads us to believe with some confidence that there are some remarkable things awaiting us based on laser technology. One thing is certain, though, this book is finished but the story of the laser has just begun.

"An early example of collaborative research with high energy density beams. Let's hope we can do better in the future!"

Index